NONLINEAR OPTICAL PROPERTIES OF ORGANIC MOLECULES AND CRYSTALS

Volume 2

This is a volume in
QUANTUM ELECTRONICS—PRINCIPLES AND APPLICATIONS

A Series of Monographs

Editors: Paul F. Liao and Paul Kelley
Founding Editor: Yoh-Han Pao (1972–1979)

The complete listing of books in this series is available from the Publisher
upon request.

NONLINEAR OPTICAL PROPERTIES OF ORGANIC MOLECULES AND CRYSTALS

Volume 2

Edited by

D. S. Chemla

AT&T Bell Laboratories
Crawford Hill Laboratory
Holmdel, New Jersey

J. Zyss

Centre National d'Etudes des Télécommunications
Laboratoire de Bagneux
Bagneux, France

1987

Published by arrangement
with AT&T

ACADEMIC PRESS, INC.
Harcourt Brace Jovanovich, Publishers
Orlando   San Diego   New York
Austin   Boston   London
Sydney   Tokyo   Toronto

ACADEMIC PRESS, INC.
Orlando, Florida 32887

United Kingdom Edition published by
ACADEMIC PRESS INC. (LONDON) LTD.
24–28 Oval Road, London NW1 7DX

Library of Congress Cataloging in Publication Data

Nonlinear optical properties of organic molecules
    and crystals.

    (Quantum electronics—principles and applications)
    Includes indexes.
    1. Molecular crystals—Optical properties.
2. Chemistry, Physical organic.   I. Chemla, D. S.
II. Zyss, J.   III. Series.
QD941.N66   1986   547.1'3   86-8070
ISBN 0–12–170612–5 (v. 2 : alk. paper)

PRINTED IN THE UNITED STATES OF AMERICA

86 87 88 89   9 8 7 6 5 4 3 2 1

Contents

Chapter III-4   Dimensionality Effects and Scaling Laws in Nonlinear Optical Susceptibilities   121
Christos Flytzanis

Chapter III-5   Trends in Calculations of Polarizabilities and Hyperpolarizabilities of Long Molecules   137
Jean-Marie André, Christian Barbier, Vincent Bodart, and Joseph Delhalle

Chapter III-6   Resonant Molecular Optics   159
B. Dick, R. M. Hochstrasser, and H. P. Trommsdorff

Contents   vii

Part III

Cubic Nonlinear Optical Effects

Chapter III-1

Basic Structural and Electronic Properties of Polydiacetylenes

M. SCHOTT

Groupe de Physique des Solides de l École Normale Supérieure
Université Paris VII—2 place Jussieu—75251 Paris 5

G. WEGNER

Max-Planck-Institut für Polymerforschung
Post Box 3148, 6500 Mainz, Federal Republic of Germany

I. INTRODUCTION

Research on polydiacetylenes is not only an integral part of modern polymer chemistry but has attracted workers from many different areas, thus creating a truly interdisciplinary field of great contemporary activity. A number of facets extending from preparative and mechanistic polymer chemistry, via quantum chemistry, spectroscopy, and materials science, to biomimetic chemistry offer many aspects to those interested in new effects and unconventional properties. The origin of all this activity is traced to a single publication (Wegner, 1969), which explained the solid-state reactivity of certain substituted diacetylenes in terms of a polymerization reaction. The

3

unique feature of this reaction is that it occurs within the perfect lattice and, being completely controlled by the packing of the monomer, leads to perfect crystals of the corresponding polymer in a number of cases. The polymerization thus proceeds as a single crystal to single crystal phase transition and has the consequence that, for the first time, macroscopic and perfect single crystals of polymers could be prepared (Wegner, 1969, 1971a,b; Kaiser *et al.*, 1972).

Polydiacetylenes exhibit a fully conjugated and planar backbone in the crystalline state and are thus considered the prototype study object with regard to the nature and physical behavior of polyconjugated macromolecules (Bloor *et al.*, 1974; Bloor, 1980, 1982).

$$nR—C\equiv C—C\equiv C—R \xrightarrow{\text{Solid state}}$$

$$R(e.g.)—(CH_2)_n—O—SO_2—\langle\!\!\!\!\!\bigcirc\!\!\!\!\!\rangle—CH_3$$

$$—(CH_2)_n—CH_3$$

$$—(CH_2)_n—O—CO—NH—Ph$$

(1)

The theoretical discussion of the electronic structure of the polydiacetylenes leads to their description in terms of a wide-band one-dimensional semiconductor (Karpfen, 1980). The optical and nonlinear optical properties of these materials are presently the subjects of intensive research, and they are treated in other chapters of this book.

A rather interesting facet of polydiacetylene research that has some impact on the development of nonlinear optical devices grew out of the successful attempt to polymerize molecules of amphiphilic character. Molecules that have the diacetylene group $—C\equiv C—C\equiv C—$ as part of the hydrophobic section of amphiphilic structures ("soaps") can be spread at the air–water interphase of a Langmuir trough and may be transferred to solid substrates by the Langmuir–Blodgett technique. Subsequent photopolymerization gives rise to thin films of molecular controlled thickness of polydiacetylenes (Tieke *et al.*, 1976, 1977). In addition to the study of their properties related to electronic and/or all-optical signal-processing devices (Sandman *et al.*, 1985), their use as resistor materials has been discussed for the fabrication of integrated electronic devices (Lando, 1985). The observation that amphiphilic diacetylenes could be photopolymerized even within a single monomolecular layer floating at the air–water interphase suggested that they could also be polymerized in the form of micellar or vesicular structures. In fact, micelles and vesicles built from diacetylenes can be permanently stabilized by photopolymerization (Gros *et al.*, 1981; Johnston *et al.*, 1980). Incorporated into suitable phospholipid structures, they can be used for reconstitution experiments with respect to certain functions of the membrane of biological cells and for testing membrane–enzyme interaction (Gros *et al.*, 1981).

The very fact that chain growth in the polymerization of diacetylenes takes place inside the perfect lattice of single crystals allows for the study of the intermediates of the reaction by a combination of electron spin resonance (ESR), nuclear magnetic resonance (NMR), and optical spectroscopy with the molecular axes oriented with respect to the probing field direction via the crystal orientation. Thus, in combination with the results from X-ray structure analyses on the monomers and corresponding polymers, the details of the polymerization mechanism could be worked out (for recent reviews, see Sixl, 1984, 1985; Schwoerer and Niederwald, 1984). The polymerization of diacetylenes is, therefore, one of the best investigated and best understood reactions in the whole realm of organic and polymer chemistry (Bloor and Chance, 1985).

II. SYNTHETIC ASPECTS AND PHYSICAL–CHEMICAL CHARACTERIZATION

A. Topochemical Polymerization as a Synthetic Tool

Solid-state polymerizations are special types of phase transitions, in which a solid monomer is transformed into a solid polymer without involvement of a liquid intermediate state. It is not surprising that the mechanism of phase change has a profound impact on the perfection of the polymer phase so produced.

The polymerization of diacetylenes is initiated by thermal treatment, or by photoradiation or high-energy radiation of the monomer crystal. It proceeds homogeneously inside the monomer crystal, starting at points distributed at random throughout the lattice (Kaiser *et al.*, 1972; Schermann *et al.*, 1975; Leyrer *et al.*, 1978; Dudley *et al.*, 1985). A solid solution first of extended-chain oligomers and later of macromolecules is formed, as schematically shown in Fig. 1(b). Consequently, the coherence of the crystal is retained and the single-crystal character is never destroyed.

This behavior is quite uncommon. Normally, solid-state reactions proceed by nucleation of a new phase at defects, surfaces, or imperfections of the parent

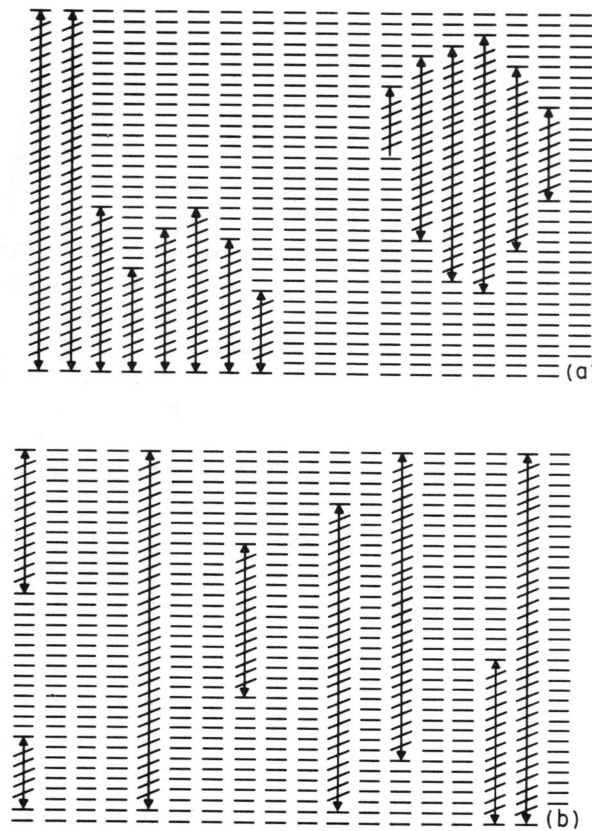

Fig. 1. Two different mechanisms of phase transformation in solid-state polymerization. (a) Heterogeneous growth by nucleation. (b) Homogeneous reaction in form of solid solution. Mechanism (b) is found in diacetylene polymerization.

phase, as indicated in Fig. 1(a). In these cases, macroscopic single crystals of the product phase cannot be obtained, because the coherence between the nucleation sites is lost with increasing conversion, due to inherent differences in density between the parent and product phases. Polycrystalline powder is usually obtained as the reaction proceeds to higher conversions. Various aspects of such considerations and their impact on the morphology and texture of the product phase as well as on the kinetics of the reaction have been treated in the recent literature (Wegner *et al.*, 1975; Wegner, 1984; Baughman and Chance, 1978; Baughman and Yee, 1978).

It must be noted that topotaxy quite often observed in solid-state syntheses—that is, coincidence of certain crystallographically defined directions of the parent and product phases—is not necessarily a consequence of topochemical effects. If nucleation occurs at the surface of a crystalline matrix, an orientation of the new phase with regard to the surface is expected because of the effect of surface free energy on the orientation of the nuclei. In contrast to topotaxy, which merely describes orientational relations between two distinct phases, the term topochemistry implies the control of reactivity, including texture and morphology, by the packing of the molecules within the lattice of the parent phase.

Furthermore, there is very little hope of devising other methods of preparation of macroscopic single crystals of long-chain molecules besides solid-state polymerization. It is a well known fact among polymer scientists that the crystallization of macromolecules almost inevitably leads to folded-chain microcrystals if the crystallization is attempted from dilute solution or from the melt of polymers (Lauritzen and Hoffman, 1960; Zachmann, 1969). All attempts to prepare a polymer backbone first by one of the methods of polymer chemistry, in the hope that the product can be crystallized in a second step to obtain the desired macroscopic single crystal texture, are highly unrealistic and doomed to fail. In addition to solid-state polymerization, true single crystals of macromolecules can only be formed by simultaneous polymerization and crystallization, but only if one succeeds in controlling the nucleation step sufficiently well (Wunderlich, 1968). Although such reactions have considerable importance in the technical production of some polymers, the nucleation processes are not well understood, so that polycrystalline materials are regularly obtained.

For the sake of clarity, the four different methods of polymer crystallization are summarized in Fig. 2.

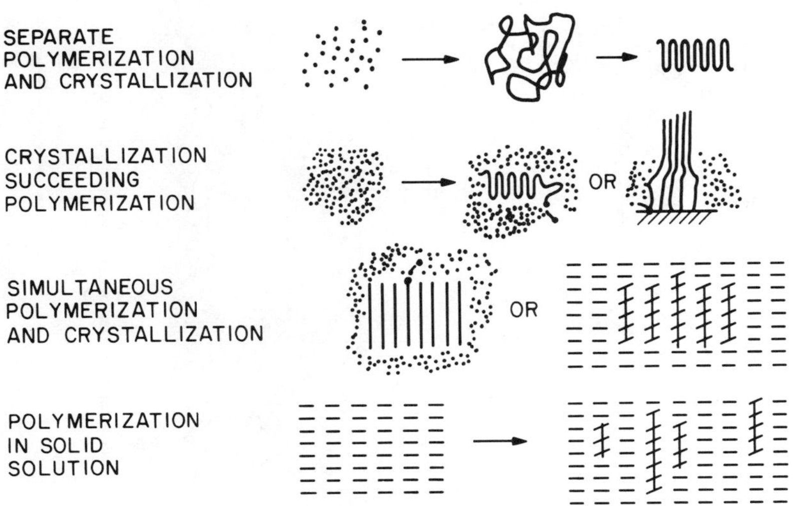

Fig. 2. Schematic representation of the four different methods for producing crystals of polymers.

In contrast to the problems encountered with crystalline and semicrystalline polymers, it is possible to prepare homogeneously oriented samples of liquid-crystalline polymers according the rules developed for normal low-molecular-weight liquid-crystal-forming compositions. This is notably true for the so-called side-chain liquid-crystal polymers (Finkelmann and Rehage, 1984).

The nature of the topochemical polymerization of diacetylenes is best explained with the help of Fig. 3. In the monomer crystal, the molecules are arranged in a ladderlike fashion such that the end of one triple-bond system approaches the beginning of the adjacent one at a distance $R \leqslant 4$ Å. Polymerization proceeds by successive tilting of each molecule along the ladder without moving the center of gravity in ideal cases. This "least motion principle" (Baughman and Yee, 1978) allows that the mode of packing of the side groups R, the specific volume, and the lattice symmetry can be retained throughout the reaction. This requires the angle ϕ between the diacetylene rod and the stacking axis to be approximately 45°. The chain repeat distance of 4.91 Å as well as the bond lengths and angles of the polymer backbone are nearly identical for all cases studied so far; relevant data are compiled in a recent review (Enkelmann, 1984).

The experimental data qualitatively confirm the geometrical model presented by Fig. 3. The reactivity is controlled by the packing of the monomer and not by the chemical nature of the substituents.

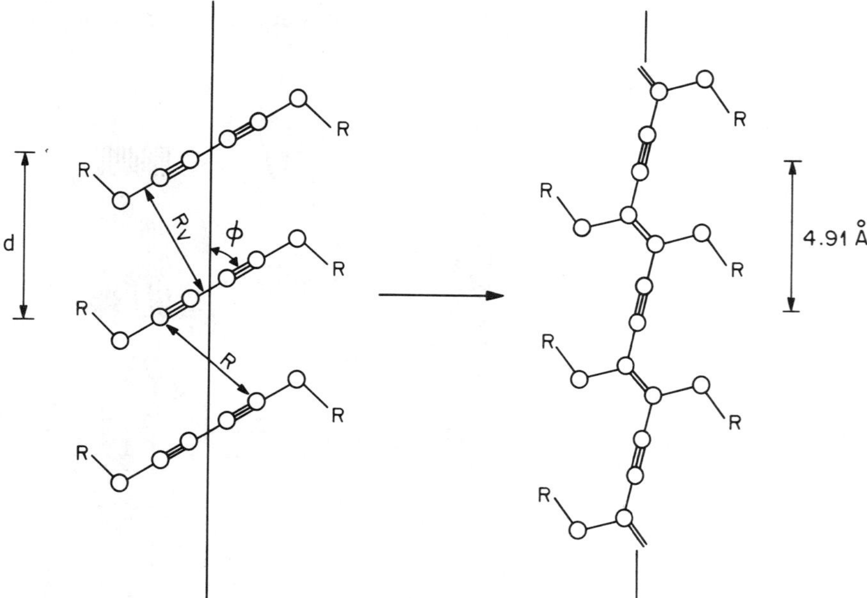

Fig. 3.   Packing requirements for solid-state polymerization of diacetylenes.

Generally, different modifications of the same monomer can be obtained and show drastically different reactivity, which is the best proof of the above statement (Kaiser *et al.*, 1972). On the other hand, the packing parameters do not provide an absolute scale for the rate of polymerization. Monomers with virtually identical packing can show large reactivity differences.

B. Scope and Limits of Chemical Design

The topochemical polymerization of diacetylenes $R-C\equiv C-C\equiv C-R$ generally allows the synthesis of polyconjugated macromolecules with the sequence single–double–triple bond, the double bonds having substituents R in trans position and being in the all-trans configuration with regard to the backbone connection, as shown in Fig. 3. In principle, all substituents R are possible that do not interfere with the required chain repeat of 4.91 Å of the polymer backbone. Thus, aliphatic or aromatic substituents are feasible as long as their own packing requirements do not force the system to adopt a structure with a lattice periodicity that is not compatible with the topochemistry of the system. Specifically, secondary or tertiary C atoms are unfavorable if attached close to the triple bond system, which is to some extent a limitation in the synthesis of optically active polydiacetylenes. Optically active polydiacetylenes can, however, be obtained, if the center of chirality is moved outwards from the reaction center. The monomers may be symmetrical $R-C\equiv C-C\equiv C-R$ or unsymmetrical $R-C\equiv C-C\equiv C-R'$, linear or cyclic. In the latter case the $-C\equiv C-C\equiv C-$ group is part of a ring. The reaction center may occur only once in the monomer, which is the usual case, but this is not a limitation. There are a number of cases known where the diacetylene group occurs twice or several times in the same molecule (Baughman and Yee, 1974). It is still unclear whether all triple bonds have the same reactivity and whether complete conversion can be reached in these unusual cases. The reactivity is not limited to the presence of substituents interacting with each other through van der Waals forces only. Excellent results have been obtained with systems exhibiting hydrogen bonding or salt structures. Specifically, the salts of diacetylenic monocarbone acids of the type $H-(CH_2)_n-C\equiv C-C\equiv C-(CH_2)_n-CO_2H$ with mono- or divalent inorganic counterions show enhanced solid-state reactivity either in bulk or in the form of Langmuir–Blodgett layers.

Although the topochemical polymerization of diacetylenes has a wide scope with regard to the chemical details of the substituents R, it is generally impossible to predict from first principles whether a given compound will undergo polymerization or not; it is even difficult, if not impossible, to extract certain rules from the very large body of experimental facts that has become available in the recent years. This reflects the fact that we totally lack a theory that would allow us to predict the packing of an organic molecule from just its molecular structure.

The attachment of a desired substituent to the polymer backbone can also be achieved by a polymer analogous reaction. In this case, a soluble polydiacetylene with reactive substituents has to be prepared first, and the desired substituent is then introduced in a subsequent reaction. Unfortunately, such reactions are rarely quantitative and, even if so, the state of a single crystal cannot be reconstituted. Nevertheless, films of glassy or polycrystalline nature can be cast from solutions of polydiacetylenes, so that this way offers some flexibility at least.

Finally, copolymers can be obtained polymerizing solid solutions of two monomers in single crystal form (Enkelmann, 1984). This is an interesting approach to form single crystals of polydiacetylenes doped with specific impurities.

The synthesis of the symmetrical monomers $R-C\equiv C-C\equiv C-R$ is carried out in a straightforward way via Glaser coupling of terminal acetylenes. This reaction is catalyzed by copper salts or complexes of copper salts.

$$2\,R-C\equiv CH \xrightarrow[\text{Cu(I)}]{\text{O}_2} R-C\equiv C-C\equiv C-R \qquad (2)$$

The synthesis of the unsymmetrically substituted diacetylenes follows a less straightforward way and sometimes suffers from poor yields or not well reproducible conditions (Tieke *et al.*, 1979).

$$R-C\equiv CI + R'-C\equiv CH \xrightarrow[\text{NH}_2\text{OH}]{\text{Cu(I)}} R-C\equiv C-C\equiv C-R' \qquad (3)$$

The polymerization of the single crystals of the monomer is achieved either thermally by anealing of the crystals 10–100°C below their melting point. The anealing procedure is usually carried out isothermally in an oven or thermostat. Protection from oxygen is usually not necessary. The time–conversion curve is in some cases linear, but in most cases S-shaped, indicating an autocatalytic behavior. The relevant experimental data are shown in Fig. 4 for the case of the *p*-toluene sulfonate of hexadiyne diol (PTS) as an example (Wegner, 1972).

Thermal polymerization is the preferred route to perfect single crystals of the polymer. Although high-energy radiation is also suitable to achieve a bulk polymerization, as demonstrated for 1,6-bis-carbazolylhexa-2,4-diyne (DCH) in Fig. 5 (Enkelmann *et al.*, 1980), it is not recommended because of the inherent danger of producing lattice defects through radiation damage. Photopolymerization is not suitable for production of large single crystals,

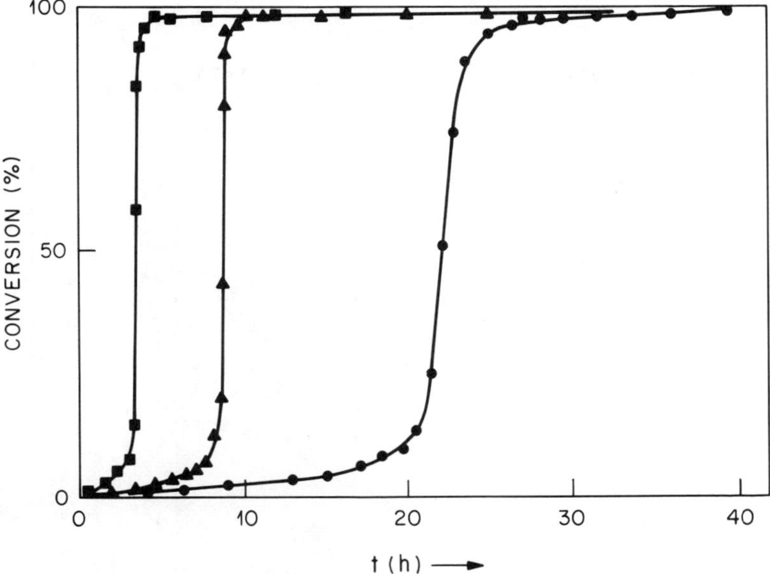

Fig. 4. Time-conversion curves for the thermal polymerization of PTS: ●, 60°C; ▲, 70°C; ■, 80°C.

since it proceeds preferably at the surface and from there into the bulk of the crystal. The small differences in density that always exist between monomer and polymer will lead to the build-up of strain between the already polymerized and the nonpolymerized part of the crystal. This will soon lead to cracks and fissures in an uncontrollable way.

C. Synthesis and Polymerization of Typical Monomers

Table I gives a compilation of relevant data concerning the synthesis and characterization of the polymerization conditions, including references to polymer properties of some of the better investigated diacetylene monomers. In addition, the detailed preparation conditions for the model polymers PTS as an example for thermal polymerization and of DCH as an example for radiation-induced polymerization will be given. The synthesis of an unsymmetrically substituted diacetylene is described for tricosa-10,12-diynoic acid as an example for a multilayer forming diacetylene. Some compounds that have been found to be suitable for the formation of multilayers as well as micellar or vesicular structure are shown in Fig. 6. Data relevant to their synthesis and biomimetic activity will be found in a recent review (Gros *et al.*, 1981).

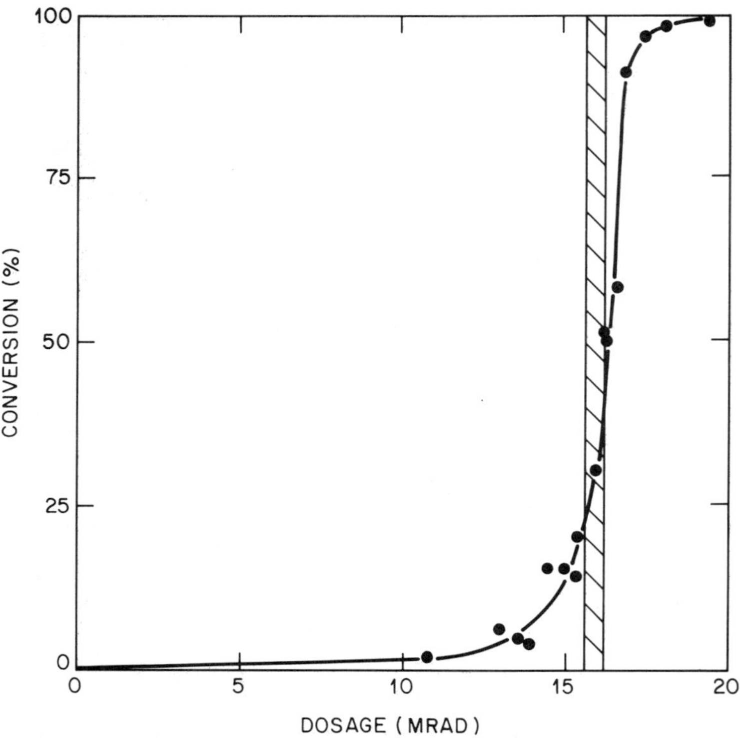

Fig. 5. Conversion versus dosage curve for the polymerization of DCH exposed to ^{60}Co γ radiation at 25°C. The shaded region indicates the occurrence of a phase transition from a less active to an active modification [From Enkelmann *et al.* (1980).]

1. Synthesis and Thermal Polymerization of Hexadiyne-1,6-diol-bis-p-toluene Sulfonate (PTS)

a. Synthesis of the Monomer (Wegner, 1971b). Eleven grams (0.1 M) 2,4-hexadiyne-1,6-diol is dissolved in 100 ml tetrahydrofuran, and 50 g tosyl chloride is added. The solution is cooled to 15°C and a solution of 20 g potassium hydroxide dissolved in 160 ml water is slowly dropped into the solution with rapid stirring. The reaction mixture is stirred for 6 hr at 25°C and is then poured into 500 ml ice water. Crystals of the desired monomer separate at once. They are filtered off and are well washed with ice water. The monomer is recrystallized from a boiling mixture of methanol (95%) and water at least three times in the dark. The monomer consists of colorless crystals, m.p. 96°C, which are very sensitive to light or heat. Usually they turn slightly red even in the course of crystallization. The monomer, especially its solutions, should be

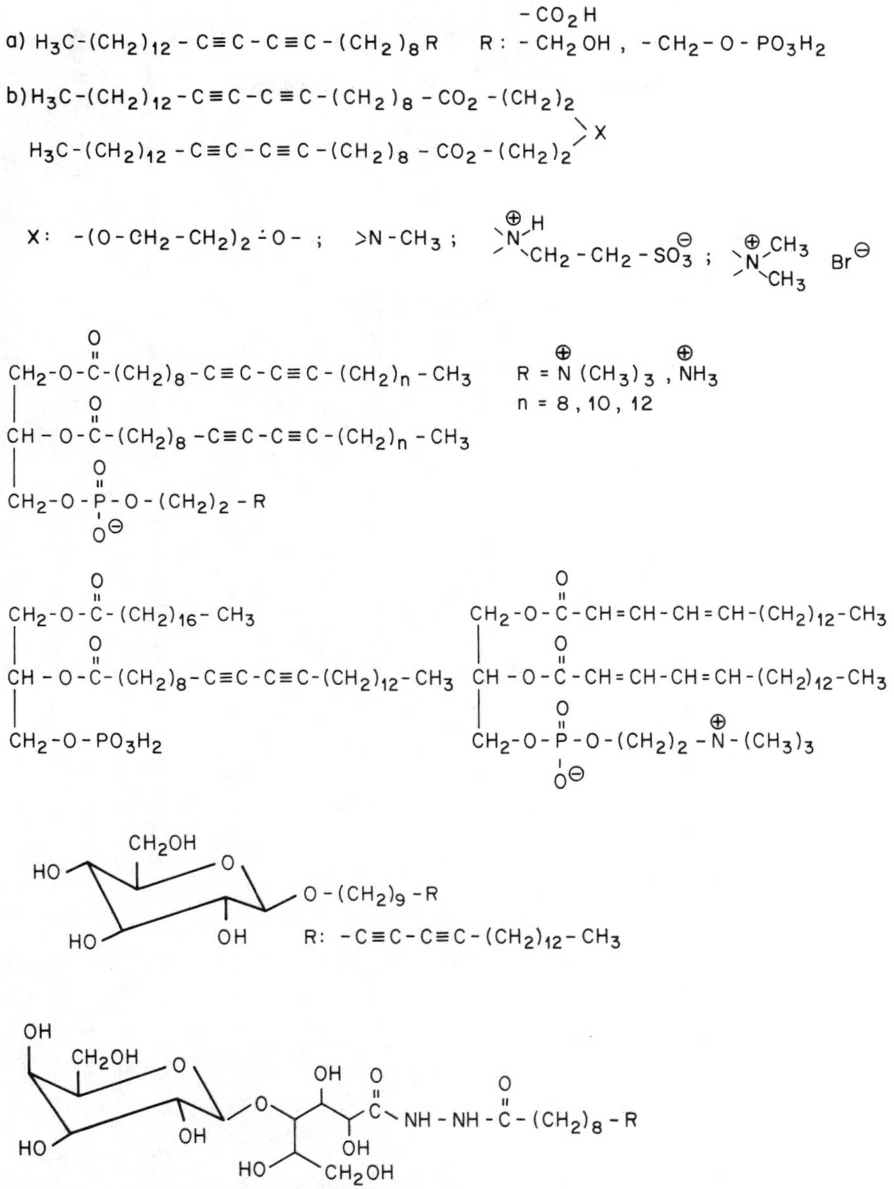

a) $H_3C-(CH_2)_{12}-C\equiv C-C\equiv C-(CH_2)_8 R$ $R: -CH_2OH$, $-CH_2-O-PO_3H_2$

b) $H_3C-(CH_2)_{12}-C\equiv C-C\equiv C-(CH_2)_8-CO_2-(CH_2)_2$
   $H_3C-(CH_2)_{12}-C\equiv C-C\equiv C-(CH_2)_8-CO_2-(CH_2)_2$ $\rangle X$

$X:$ $-(O-CH_2-CH_2)_2-O-$; $\rangle N-CH_3$; $N^{\oplus}(H)-CH_2-CH_2-SO_3^{\ominus}$; $N^{\oplus}(CH_3)(CH_3)$ $Br^{\ominus}$

$CH_2-O-\overset{O}{\overset{\|}{C}}-(CH_2)_8-C\equiv C-C\equiv C-(CH_2)_n-CH_3$ $R = N^{\oplus}(CH_3)_3$, $NH_3^{\oplus}$
$CH-O-\overset{O}{\overset{\|}{C}}-(CH_2)_8-C\equiv C-C\equiv C-(CH_2)_n-CH_3$ $n = 8, 10, 12$
$CH_2-O-\overset{O}{\overset{\|}{P}}-O-(CH_2)_2-R$
         $O^{\ominus}$

$CH_2-O-\overset{O}{\overset{\|}{C}}-(CH_2)_{16}-CH_3$
$CH-O-\overset{O}{\overset{\|}{C}}-(CH_2)_8-C\equiv C-C\equiv C-(CH_2)_{12}-CH_3$
$CH_2-O-PO_3H_2$

$CH_2-O-\overset{O}{\overset{\|}{C}}-CH=CH-CH=CH-(CH_2)_{12}-CH_3$
$CH-O-\overset{O}{\overset{\|}{C}}-CH=CH-CH=CH-(CH_2)_{12}-CH_3$
$CH_2-O-\overset{O}{\overset{\|}{P}}-O-(CH_2)_2-N^{\oplus}-(CH_3)_3$
         $O^{\ominus}$

$O-(CH_2)_9-R$ $R: -C\equiv C-C\equiv C-(CH_2)_{12}-CH_3$

$\overset{O}{\overset{\|}{C}}-NH-NH-\overset{O}{\overset{\|}{C}}-(CH_2)_8-R$

Fig. 6. Examples of amphiphilic diacetylene monomers that are able to be polymerized in micellar or vesicular structures (Gros *et al.*, 1981; Bader *et al.*, 1985).

TABLE I

Representative Examples for Diacetylene Monomers, Polymerization Behavior, and Physical Properties of the Polymer, with General Monomer Structure R—C≡C—C≡C—R′

Formula	Common abbreviation	Representative literature reference for synthesis	Polymerization	Polymer properties
R = R′ —CH_2—OSO_2—〈benzene〉—CH_3	PTS	Wegner (1971b); Bloor *et al.* (1975)	Wegner (1971b); Garito *et al.* (1979); Enkelmann *et al.* (1979); Leyrer and Wegner (1979); Niederwald and Schwoerer (1983)	Wegner (1971b, 1972); Bloor *et al.* (1975); Bloor and Preston (1976, 1977); Enkelmann *et al.* (1979)
—CD_2—OSO_2—〈benzene〉—CH_3 and other selectively deuterated or ^{13}C-marked isomers	—	Kröhnke *et al.* (1980)	Bässler (1984); Hädicke *et al.* (1971)	
—CH_2—OCONH—〈benzene〉	HDU	Wegner (1969)	Wegner (1969, 1972)	Wegner (1969); Hädicke *et al.* (1971)
—$(CH_2)_4$—OSO_2—〈benzene〉—CH_3	PTS-12	Wenz and Wegner (1982)	Siegel *et al.* (1982)	Wenz and Wegner (1982); Siegel *et al.* (1982); Wenz *et al.* (1984)

Structure		Abbreviation			
$-(CH_2)_3-OCONHCH_2CO_2-n\text{-}C_4H_9$		3-BCMU	Patel (1978a)	Wenz *et al.* (1984); Patel (1978a)	Wenz *et al.* (1984); Patel (1978a); Patel *et al.* (1979); Lim *et al.* (1985)
carbazole: $-CH_2-N\!\!<$		DCH	Enkelmann *et al.* (1977); Apgar and Yee (1978)	Enkelmann (1984); Enkelmann *et al.* (1977)	Apgar and Yee (1978); Enkelmann *et al.* (1980); Young (1985)
phenyl–$NHCOCH_3$		DAAD	Wegner (1971a)	Wegner (1971a)	Enkelmann (1984)
$-CH_2$–phenyl(NO_2)(O_2N)		DND	McGhie *et al.* (1981)	McGhie *et al.* (1981)	McGhie *et al.* (1981)
$R \neq R'^{a}$ $R = -(CH_2)_m-CH_3$ $R' = -(CH_2)_n-CO_2H$		*m/n* Acids	Tieke *et al.* (1979)	Tieke *et al.* (1977, 1979); Bubeck *et al.* (1982)	Lieser *et al.* (1980)

[a] Monomers suitable for LB film formation and subsequent polymerization.

handled with care, since it is a powerful skin irritant to persons of allergic response.

b. Thermal Polymerization. In order to obtain large single crystals of up to several centimeters in edge length, the monomer is dissolved in acetone to prepare a saturated solution at room temperature. The solution is filtered into a wide-bottom Erlenmayer flask and put into a dark dessiccator. Crystallization from this solution is controlled by slow evaporation of the acetone under slightly reduced pressure, so that all of the solvent is distilled off in the period of approximately 1 week. It is important to carry out the crystallization under water-free conditions in order to prevent impure crystals from being formed.

The crystals are placed into a thermostat set at the desired reaction temperature (see Fig. 4 for the time–conversion curves). Quantitative yield is typically obtained after 12 hr of annealing time at 70°C.

Partially converted material can be extracted from residual monomer by acetone. Polymer obtained at partial conversion up to 50% is of lower molecular weight and can be dissolved by gentle heating in nitrobenzene, γ-butyrolactone, dimethylformamide, or similar polar solvents.

2. Synthesis and Polymerization of 1,6-Di-(N-carbazolyl) 2,4-Hexadiyne (DCH)

DCH is synthesized by oxidative coupling of N-(2-propinyl) carbazole (Yee and Chance, 1978; Enkelmann *et al.*, 1977). The latter is obtained from the reaction of 1-bromopropyne with carbazolyl sodium in liquid ammonia (Chadiot and Chodkiewidcz, 1969).

$$(4)$$

a. N-(2-Propinylcarbazol). Carbazole [16.7 g (0.1 M)] is dispersed in 600 ml liquid NH_3 at $-40°C$ in which previously 0.11 M $NaNH_2$ had been dissolved. The mixture is stirred for 2.5 hr and then 16.6 g (0.14 M) of propargyl bromide are added in small portions over a period of 0.5 hr. The temperature is maintained at $-55°C$ with stirring for 5 hr. NH_3 is then distilled off and the residue is extracted with hexane. The desired product crystallizes in the form of colorless crystals, m.p. 110°C, yield 17.4 g (85%).

b. 1,6-Di-N-carbazolyl-2,4-hexadiyne, General Procedure of Coupling of Terminal Acetylenes (Hay, 1962). CuCl (14 mmol) is dispersed in 100 ml solvent (acetone, dimethylformamide, or dioxane is suitable) and 14 mmol N,N,N',N'-tetramethylethylene diamine are added. A solution of 20 mmol of the terminal acetylene in 50 ml solvent is added. Oxygen is now bubbled through the reaction mixture under vigorous stirring for at least 30 min, usually for several hours to bring the reaction to completion. The temperature is kept between 25 and 35°C.

The reaction mixture is poured into water, from which the product separates in the form of crystals. It is washed with dilute HCl to remove the last traces of copper salts and then recrystallized.

DCH is prepared best in dimethylformamide as the solvent, yield 68%, colorless needles, m.p. > 270°C, with concurrent thermal polymerization. DCH is recrystallized from dimethylformamide by slow evaporation, and large single crystals of the size $0.1 \times 0.2 \times 5$ cm can be obtained without seeding.

Polymerization to perfect crystals of the polymer is achieved exposing the crystals to ^{60}Co γ-radiation at room temperature. Quantitative yield is reached at a dosage of minimum 20 Mrad. Thermal polymerization does not lead to single crystals but to polycrystalline samples, because of an intermediate phase change (Enkelmann, 1984).

3. Synthesis of Tricosa-10,12-diynoic Acid (Tieke et al., 1976)

$$H_3C-(CH_2)_9-C\equiv C-I + HC\equiv C-(CH_2)_8-CO_2H$$

$$\xrightarrow[\text{Cu(I)}]{\text{H}_2\text{NOH}} H_3C-(CH_2)_9-C\equiv C-C\equiv C-(CH_2)_8-CO_2H \tag{5}$$

Undecynoic acid (0.02 M) is neutralized with 10% KOH (15 ml). To this solution are added successively hydroxylamine hydrochloride (100 mg) and copper(I) chloride (500 mg) in 70% aqueous ethylamine (4 g). A solution of 1-iodo-1-dodecyne (0.02 M) in methanol (10 ml) is then added dropwise with vigorous stirring to the cold reaction mixture. After acidification with $2N$ H_2SO_4, the mixture is extracted with ether, the ethereal phase is dried over sodium sulfate, and ether is removed by distillation. The residue is treated with petroleum ether (b.p. 40–60°C), from which the desired monomer crystallizes on cooling as colorless platelets, yield 3.5 g (45.5%), m.p. 56.5°C.

a. Preparation of Multilayers (Lieser *et al.*, 1980).   Tricosa-10,12-diynoic acid is dissolved in chloroform (1 g/l) and spread at the surface of a film balance (MGW Lauda) working as a Langmuir trough. The water of the subphase is triply distilled prior to the filling of the trough. The pH is 6.2–6.3, $CdCl_2$ is added to the subphase to adjust a concentration of 10^{-3} mol/l. Transfer of the monomolecular layer is carried out at a surface pressure of 20 mN m^{-1} at a subphase temperature of 12°C.

4. Photochemical Stability of Polydiacetylenes

Polydiacetylenes are generally poorly soluble in organic solvents. In recent years, however, an increasing number of representatives of this class of polymers have been synthesized, showing excellent solubility in many different organic solvents due to enhanced side-chain–solvent interaction. Solutions of PDA suffer photodegradation if irradiated by ultraviolet (UV) light (Wenz and Wegner, 1982). It was shown that the photodegradation of poly[1,2-bis[4-(*p*-tolylsulfonyloxy)butyl]-1-buten-3-ynylene] (PTS-12) gives rise to a normal distribution of the molecular weight and may thus be used to prepare samples in a deliberately chosen molecular-weight range from the initial pristine polymer, which is generally of very high molecular weight. This in turn allows for the study of the molecular-weight dependence of a number of important properties that characterize the nature and structural behavior of dissolved PDAs, such as the radius of gyration or the hydrodynamic radius (Wenz *et al.*, 1984).

A dilute solution of P-3BCMU (poly-3BCMU; see Table I) suffers rapid loss of its initial reduced viscosity $(\eta_{sp}/c)_0$ if irradiated with unfiltered light from a high-pressure Hg lamp, where c is the concentration. The time dependence of η_{sp}/c at constant irradiation conditions suggests a random chain scission mechanism without depolymerization and is not indicative of a photodepolymerization, which could probably be expected as the reverse of the photochemically induced and topochemically controlled solid-state polymerization. It is also worth mentioning that photocross-linking and/or gel formation, which is common for most unsaturated polymers, has never been observed until now for any of the soluble polydiacetylenes in either their concentrated solutions or their solvent cast-films. The same degradation behavior as prevailing in dilute solutions is observed under these conditions, even for the bulk material or polymer, in the form of Langmuir–Blodgett (LB) layers.

It is concluded (Müller and Wegner, 1984) that photodegradation does not arise by direct optical excitation of the polymer backbone, but is rather a secondary result of the photoexcitation of the side groups and/or other components of the solution. This notion is supported by the finding that the rate of degradation under otherwise constant conditions may be enhanced by addition of, for example, benzophenone, anthraquinone, or 2,2'-azoiso-butyronitrile (AIBN). The presence of oxygen enhances the rate as well. The degradation is quenched by the addition of typical free-radical scavengers, such as 2,5-di-*tert*-butyl phenol or triethylamine.

This result indicates a radical mechanism for the degradation reaction, especially since the effect of AIBN can only be understood as a consequence of its ability to undergo rapid photodecomposition with concomitant production of free radicals. Consequently, degradation of the PDA chain also occurs in the dark at elevated temperatures, if suitable amounts of thermo-labile radical donors are present.

In order to explain the chain scissions, the following mechanism is proposed: A radical X· will attack the trans double bond of the polymer backbone. A short-lived intermediate will be formed, which rapidly undergoes homolysis. Two new chain ends of free-radical and carbene character will be obtained, which will be unstable in the presence of excess solvent and will rapidly decay by insertion, abstraction, or combination reactions, so that saturated chain ends are the final results of the reaction sequence shown below.

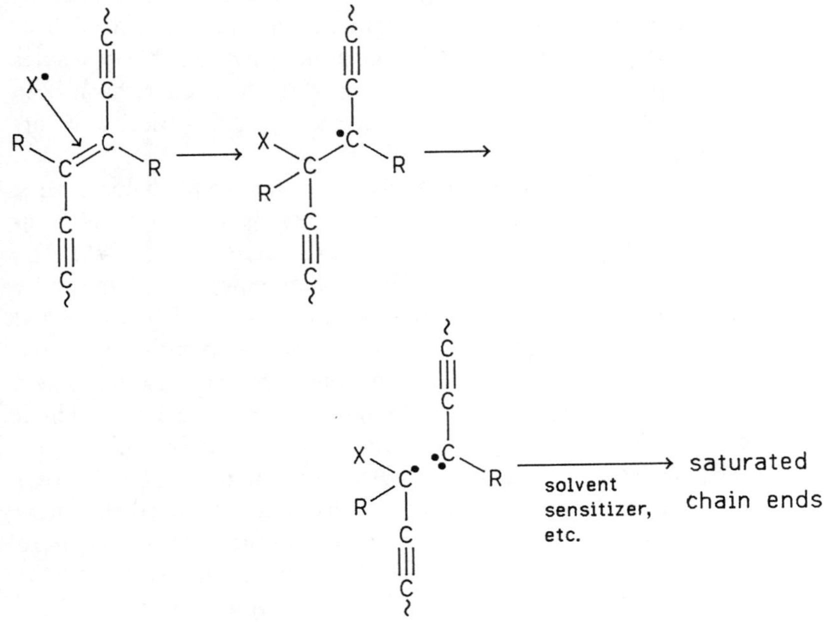

$$
\text{(6)}
$$

Although this mechanism is still hypothetical and the postulated carbene species has not been observed directly, it explains nicely the otherwise surprising effect that PDAs, despite their polyconjugated structure, do not spontaneously crosslink, but rather degrade under conditions where a sample of polyisoprene, polybutadiene, or similar polymers would rapidly show extensive cross-linking.

III. ELECTRONIC PROPERTIES

A. Introduction

A spectacular consequence of diacetylene polymerization is an enormous change in the optical properties of the polymerizing crystal. The lowest-lying electronic excited state of the diacetylene moiety is an optically forbidden triplet at an energy close to 3 eV (Bertault *et al.*, 1979). Singlet states excitation begins around 4 eV (see, for instance, for HDD, Kawaoka, 1976), so photopolymerization requires photons with $\lambda < 320$ nm in the absence of sensitization. Most side groups considered to date also absorb in the UV only, either above (in PTS or BCMU, for instance) or below (in DCH for instance) diacetylene itself, so the corresponding crystals and films are transparent. Actually, reactive monomer solids are often colored blue, pink, or yellow-orange by the few polymer chains produced by almost unavoidable exposures to heat or radiation. A few diacetylenes, some of them reactive, have been made with chromophoric side groups, thus generating a rich variety of colors (Patel, 1981).

Polymerization turns these transparent materials into deeply colored ones, often with a golden metallic reflectance, and increases the optical non-linearities by up to three orders of magnitude (Sauteret *et al.*, 1976). The absorption is highly polarized: rotation of light polarization will change it by two orders of magnitude. These properties are those of a highly conjugated polymer backbone with extensive electron delocalization. A polydiacetylene crystal contains perfectly organized, very long, one-dimensional unbranched chains. This is visually shown by the direct imaging of the crystal lattice at high magnification in an electron microscope (see Young, 1985). Such chains are weakly coupled electron systems that possess many interesting (and poorly understood) properties. Nonlinear optical properties are discussed elsewhere in this book. Here, we shall concentrate on a few questions directly relevant to the study of these properties: what are the electronic ground state and the nature of the excited state responsible for the (linear) optical properties? We shall discuss photoconductivity only to the extent that it gives information on the excited states, and leave aside any consideration of transport properties. Similarly, we shall almost not discuss partially polymerized systems, and concentrate on pure polymers, or crystals containing the maximum polymer concentration that can be achieved. However, we shall briefly touch on the properties of disordered diacetylenes, mainly in solution.

Polydiacetylene crystals, however, are not made of conjugated polymer chains only. Most of the atoms in the unit cell usually belong to side groups attached to the chain; so the properties like infrared (IR) absorption, or even lattice phonon dispersion, where the chain atoms are not favored, are dominated by side groups. Chemical degradation probably starts on side groups, as suggested by the high thermal stability (up to 300°C) of DCH compared to the lability of PTS, where decomposition is detected down to 60°C (Yee and Chance, 1978). Electronically, the chains might be thought to be well decoupled from the side groups, even those that contain conjugated structures like benzene, pyridine, or carbazole, but this requires qualification. No polydiacetylenes have been made with side groups in strong interaction with the chain—for instance through charge transfer, as in TCNQ salts. In the presently known polymers, conjugated rings are almost electronically decoupled from the chain by one or several CH_2 linkages, necessary to meet the requirement of crystal polymerization (see above). In the crystal, however, a side group can be quite close to another chain, as in PTS for instance (Kobelt and Paulus, 1974; Aimé *et al.*, 1982). As we shall see below, chain–side-group interactions affect chain transition energies, and Raman frequencies through Fermi resonances (Batchelder and Bloor, 1979).

B. Ground-State Geometry of the Chain

1. Introduction

Unlike polyacetylene, a polydiacetylene must have C—C bonds differing in lengths, and its ground state is nondegenerate. The π electrons can be moved along the carbon skeleton; two structures that can be written with ordinary C—C bonds are shown in Fig. 7. They have been shown very often in the literature and called butatrienic (**1a**) and acetylenic (**1b**). For the latter, the name enyne seems more appropriate. The geometries are different, as well as the energies (except possibly by accident), so a polydiacetylene chain will not support a soliton in the polyacetylene sense.

The question then is: what is actually the ground state of the chain: butatriene, enyne, or some "admixture" of the two? Although the pure states differ in energy, this "admixture" may be caused by a suitable "crystal field" perturbation, for instance. Since the study of optical absorption (to be studied in Section III,C) indicated the existence of two contrasting absorption spectra with different vibrational frequencies, it was assumed early on that they corresponded, more or less, to the two limiting structures of Fig. 7, and

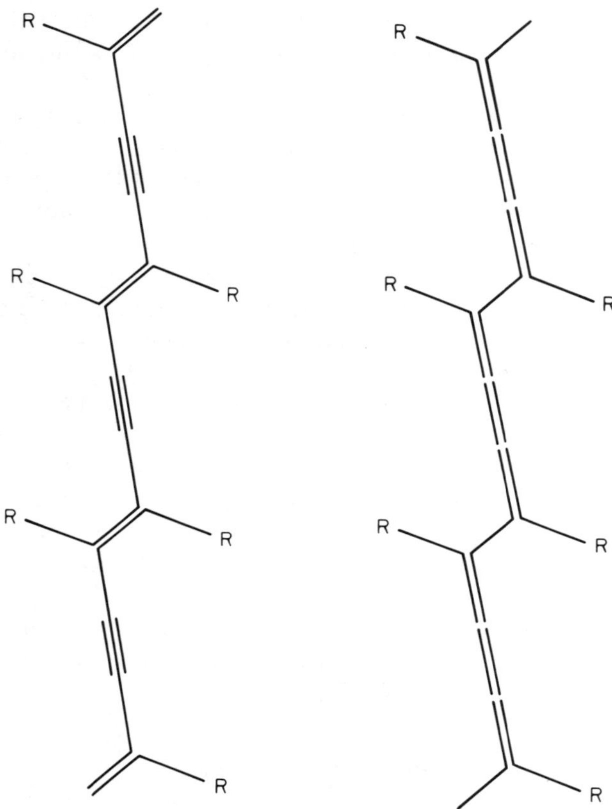

Fig. 7. Two proposed electronic configurations of polydiacetylene chains: the enyne structure is shown at left, the butatriene one at right; R, side groups. Typical side-group molecular structures are given in Table III.

crystallographic evidence was cited in support. Now, the mood is changing; but mood is not necessarily truth. So we shall first try to review the available knowledge on the structure of polydiacetylene ground chain in polymer single crystals. Then we shall turn to chains in solution in Section III,E. We shall not discuss the structure of polymer chains dispersed in monomer crystals.

2. Structural Studies of Polymer Chain Geometry

In a single crystal, the exact ground-state molecular geometry can be determined crystallographically, and compared to model enyne or butatriene molecules. Table II shows the geometry of a few such model compounds; the differences are well beyond crystallographic accuracy.

TABLE II

Geometry of Model Compounds[a]

Compound	Structure			Source
Experimental Butene-1-yne-3 (vinylacetylene)	1.342	1.431	1.209	Fukuyama *et al.* (1969)
Hexadiene-1,5-yne-3 (divinylacetylene)	1.347	1.425	1.220	Almenningen *et al.* (1984)
Butatriene	1.318	1.283		Almenningen *et al.* (1961)
Tetraphenyl butatriene	1.348	1.260	1.478	Berkovitch–Yellin and Leiserowitz (1977)
	1.332	1.261		Irngartinger and Jäger (1976)
Theoretical bond lengths[b] Enyne	1.321	1.425	1.194	Karpfen (1980)
Butatriene	1.444	1.319	1.248	

[a] All bonds lengths are in angstroms.
[b] Bond lengths indicated are those obtained with the 7s3p/3s basis set.

The geometries of several polydiacetylenes, as published in the literature, are given in Table III; what safe conclusions do they allow?

In a few favorable cases, a good monomer single crystal can be grown, and completely polymerized into a good single crystal, without going through a phase transition, and in comparatively "mild" conditions (for instance annealing around 60°C), avoiding creating too much physical and chemical disorder. The standard case is PTS, in which the room-temperature crystal structures of monomer (Aimé *et al.*, 1982) and polymer (Kobelt and Paulus, 1974; Cottle, 1980) are accurately known. PTS is clearly an enyne polymer, and the two published structures agree very well. Another favorable case

TABLE III

Bond Lengths in Diacetylene Polymers[a]

Usual name	Side-group formula	Ref.[b]	Bond Lengths in Å			Number of reflections	Final R	Remarks[c]
			A	B	C			
PTS	$-CH_2OSO_2\phi CH_3$	a	1.191 (4)	1.428 (4)	1.356 (4)	1700	0.055	1
		b	1.195 (5)	1.424 (3)	1.364 (5)			
THD	$-CH_2N\phi_2$	c	1.205 (2)	1.426 (3)	1.359 (3)	3800	0.068	2
HDU	$-CH_2OCONH\phi$	d	1.21	1.41	1.36	2000	0.079	3
DCH	$-CH_2$-carbazole	e	1.21	1.44	1.33	425	0.086	4
		f	1.23 (1)	1.42 (1)	1.38 (1)	945	0.098	
MBS	$-CH_2OSO_2\phi OCH_3$	g	1.22 (2)	1.40 (2)	1.27 (3)	?	0.13	5
PFBS	$-CH_2OSO_2\phi F$	h	1.17 (2)	1.39 (3)	1.42 (3)	472	0.069	
PTS-12	$-(CH_2)_4OSO_2\phi CH_3$	i	1.15 (2)	1.41 (1)	1.41 (2)	1050	0.098	6
TCDU	$-(CH_2)_4OCONH\phi$	j	1.17 (2)	1.38 (2)	1.46 (2)	728	0.093	7
DNP	$-CH_2O\phi(NO_2)_2$	k	1.205 (7)	1.410 (5)	1.471 (9)			8
BPG	$-\phi OCO(CH_2)_3COO\phi-$	l	1.29	1.38	1.42	902	0.075	9
4-PBrPU	$-(CH_2)_4OCONH\phi Br$	m	1.16	1.45	1.36	1340	0.08	10

[a] The meaning of bonds A, B, and C is indicated on Table II. Technical details were not available for refs. b and k. Several monomer modifications, of varying reactivities, are known at least for HDU, MBS, and TCDU.

[b] References: (a) Kobelt and Paulus (1974); (b) Cottle (1980); (c) Enkelmann and Schleier (1980); (d) Hädicke *et al.* (1971); (e) Apgar and Yee (1978); (f) Enkelmann (1984); (g) Williams *et al.* (1980); (h) Enkelmann (1984); (i) Siegal *et al.* (1982); (j) Enkelmann and Lando (1978); (k) McGhie *et al.* (1981); (l) Day and Lando (1978); (m) Spinat *et al.* (1985).

[c] Remarks: (1) thermally polymerized at 60°C; (2) thermally polymerized at 70°C; (3) UV photopolymerization, possibly inhomogeneous; (4) strongly first-order phase transition around 25–30% polymer content, with a 5% volume change (Enkelmann *et al.*, 1980); (5) in the review by Enkelmann (1984), MBS is erroneously given as a typical enyne polymer. In fact, the published structure (Williams *et al.*, 1980) is of low accuracy and does not agree well with an enyne structure; this work too used an incompletely polymerized crystal. (6) A phase transition with reduction of the unit cell size occurs during polymerization, and the motion of the diacetylene during the reaction is very large. (7) The investigated crystal was not completely polymerized and solvent extraction was used to remove ~10% residual monomer (Iqbal *et al.*, 1977); (8) A phase transition disorders the crystal before complete polymerization occurs, so this sample was probably not 100% polymer. (9) The monomer is cyclic. The maximum polymer content achievable is 35%. Thus, the structure is that of a monomer–polymer mixed crystal in which only a few atoms were given different positions in monomer and polymer. The polymer being the minor constituent, the corresponding bond lengths are the less reliable. (10) Heterogeneous polymerization. Structure solved at 205 K. The authors mention that at 295 K, $A = 1.20$ Å. Such a temperature variation is surprising.

seems to be THD (Enkelmann and Schleier, 1980), where the polymer geometry is undistinguishable from that of PTS—monomer THD structure is not known. Yet another example may be the 2:1 complex of HDU with dioxane, the very first polydiacetylene whose structure was determined (Hädicke *et al.*, 1971) and found to be an enyne one; however, it was photopolymerized, which may have lead to inhomogeneity in the polymerization process. Therefore, the occurrence of the enyne structure is well established in polydiacetylenes.

Other structures given in Table III show large variations, and the geometry of TCDU (Enkelmann and Lando, 1978), in particular, has been claimed repeatedly in the literature to demonstrate the occurrence of a butatrienic structure. In fact, in all cases the R factor is poor, considering the small number of reflections compared to the fairly large number of independent atoms in the unit cell. In addition, there are reasons to doubt the quality of the polymers: polymerization may be incomplete (MBS, TCDU, and possibly DNP), or a phase transition, with large unit cell and/or atomic position changes (DCH, PTS-12), may occur during polymerization. It is difficult to understand how a C—C bond can be significantly shorter than a triple bond, as claimed in PTS-12. Of course, some unspecified "disorder" might be invoked. Several known effects can create bond-length errors. Rotational oscillations of molecules, if not properly accounted for, may reduce the calculated bond length (Cruickshank, 1956; Busing and Levy, 1964). Another more likely cause is the influence of residual monomer clusters: reacting carbons move by more than 1 Å from monomer to polymer position within the unit cell; residual monomer carbons will distort the electron-density map, and this will show up in the results of most fitting programs as an anisotropic thermal factor and slightly shifted average atomic position. An apparent shift of 0.05 Å for such carbons is enough to make the double bond of the enyne structure look like a single bond, as observed. It would be useful to assess how much residual monomer would be needed to do this. Note that the absence of monomer Bragg reflections, sometimes quoted as proof of complete polymerization, only shows that there is no macroscopic polymer–monomer phase separation, and gives no indication about monomer concentration in a mixed monomer–polymer crystal, where only an average, concentration-dependent structure, is observed (Enkelmann, 1984; Aimé, 1983). Summarizing, the anomalously short triple bond suggests systematic errors. The similarity of several structures (PFBS, PTS-12, TCDU, DNP) may correspond to the existence of a common polymer geometry, different from the enyne one, but may as well reflect a common technical difficulty.

The case of BPG (Day and Lando, 1978) is very special, since only a mixed polymer–monomer crystal, containing only one-third polymer, can be studied. In addition, the polymer backbone is highly strained, not even planar, and the chain length is unknown and might be quite short. So this material should not, in fact, be taken into account in our discussion of pure-polymer chain geometry.

The only firm conclusion allowed by crystallography of polydiacetylenes to date is the occurrence of the enyne polymer structure. Significantly different geometries have been claimed, but they cannot be deduced from experiment at present; it may even be that all crystalline polydiacetylenes investigated to date are enyne polymers. Butatrienic polymers may exist, but their existence is not proven. It is certainly too early to speculate from these data about the possible influence of side groups on polydiacetylene configuration. Further work is surely needed.

Another conclusion is that the crystal quality of polydiacetylene crystals is rarely excellent, and that one should not forget the possible presence of unreacted monomer. Macroscopically good crystals of PTS and PFBS can be obtained. The same has been claimed for DCH crystals, although the crystal goes through a first-order transition during polymerization (Enkelmann, 1984).

3. Theoretical Studies of the Polydiacetylene Ground State

The ground-state geometry can also be calculated theoretically by energy minimization of a one-dimensional chain with simple side groups (H or CH_3). This was done by Karpfen (1980), where reference to earlier work may also be found. Several of his conclusions are of interest. First, the enyne structure is the more stable, the energy difference with the butatriene structure being ~ 0.5 eV per repeat unit. Second, the calculated enyne geometry is in fairly good agreement with the one observed for PTS or THD; the bond lengths may be a trifle too short, but this may be a general weakness of the calculation, since the theoretical bond lengths for butene-1-yne-3 are also shorter than the observed one (see Table II). Third, the calculated butatrienic structure does not look like any structure in Table III, with the possible exception of the BPG mixed monomer–polymer crystal. Finally, the potential minimum corresponding to the butatrienic structure is very flat; there might even be no minimum at all, since the potential was only calculated around one isomerization route, which may not correspond to the steepest descent.

On the basis of this calculation, one would then expect that all polydiacetylenes would be enynes, except when the environment provides an energy larger than 0.5 eV per monomer in favor of the butatriene; since the calculated repeat lengths of the two geometries differ by 2% only, and can be adjusted by bond angle changes that would not greatly change the energy, compression or stretching of the chain alone would not easily provide this energy difference. Some more direct electronic effect would seem to be needed.

We cannot discuss in detail here all the other calculations. They vary technically a great deal, including or not electron correlations, considering one-electron excitations or excitons, and so on. A general feature is the greater stability of the enyne structure (see, for instance, Whangbo *et al.*, 1979; Bredas *et al.*, 1981; Cade and Movaghar, 1982), often by a large amount: 0.5 eV in the calculation of Bredas *et al.*

The same authors find that the ordering of the highest occupied π bands differs in the enyne and the butatriene structures: at $\mathbf{k} = 0$, the symmetry of the highest band is A_u in the former case, B_g in the latter. When the band structure is calculated using the published TCDU chain geometry (Enkelmann and Lando, 1978)—which, as discussed above, may be in doubt—the corresponding symmetry is again A_u, so that, in that sense, one might say that even the TCDU structure is, electronically speaking, enyne, although its ground-state energy is barely below that of the butatrienic (by less than 0.1 eV), again indicating how shallow the butatrienic minimum is, if it exists at all.

4. Experimental Evidence from Oligomers

The intermediate steps of low-temperature photopolymerization have been studied in several diacetylene monomer crystals (for a review, see Sixl, 1984). In all cases, short intermediates are diradicals, which implies a butatriene structure, and they turn into dicarbenes approximately when the heptamer is formed. Longer intermediates are dicarbenes, which implies an enyne structure. Deactivation of one or both chain ends yields asymmetric carbenes and stable oligomers, all with enyne chain structure. This is found in several cases where the corresponding polymer departs from the pure enyne geometry, according to Table III: TCDU itself, PTS-12, or PFBS. Even BPG (Bubeck *et al.*, 1979) shows a carbene ESR signal in a crystal containing a few percent polymer.

It would be surprising—but cannot be excluded, since internal strain in the crystal varies during the polymerization—to discover that these materials revert to a butatrienic structure on further reaction.

C. Optical Properties of Polydiacetylene Crystals

1. Different Types of Visible Absorption

Although there is no proof of different ground-state geometries, different absorption spectra have been observed, or rather reflection spectra, a few of which on a large enough frequency range to allow Kramers–Kronig inversion. The position of the first, lowest energy, absorption, or reflection peak of several polydiacetylene crystals at room temperature is given in Table IV.

Comparison of Table IV with Table III shows no simple relation between C—C bond lengths or polymer repeat distances and transition energies. Still, Table III clearly shows that there are two contrasted types of absorption spectra: those peaking at $15{,}800 \pm 500$ cm^{-1} and that show quite a lot of vibronic structure (Fig. 8), and those peaking around $18{,}500-19{,}000$ cm^{-1} and are much less structured (Fig. 9). In all cases, the absorption or reflection is highly polarized along the polymer chain direction. Figure 8 shows the reflection of DCH for light polarized perpendicular to the chains: it is smaller by an order of magnitude. The corresponding absorption would be $\sim$100-fold smaller.

Often in the literature has a spectrum of the first type been associated to enyne structures, the prototype being PTS, and of the second to butatriene structures, with TCDU phase I as a prototype; however, as we have seen, the crystallographic evidence is faint. The new fashion of the day would be to associate these differences with "strain on the backbone." As such, this is an ill-defined concept. One way to test it would be to study the effect of known applied strain, or stress. There have been a few such experiments, which we discuss in Section III,C,3.

TABLE IV

First Optical Absorption Peak Wavelength (cm^{-1}) of Polydiacetylenes at Room Temperature

PDA	KK Inversion peak	Reflection peak[a]	References
PTS	16,100		Bloor and Preston (1976)
	2 K: 15,650 and 15,910		
DCH	15,300[b]		Hood *et al.* (1978)
	2 K: 15,000		Sebastian and Weiser (1981a)
PFBS		16,200	Chance *et al.* (1979)
IPUDO		15,400	Eckhardt *et al.* (1979)
		130°C: 18,650	
4-BCMU		15,800	Chance *et al.* (1979)
DNP		18,800	Albouy (1982)
TCDU			
Phase I[c]		18,470	Tokura *et al.* (1984)
		18,800	Müller *et al.* (1977)
		10 K: 18,070	Sebastian (1980)
Phase II		$\sim$15,300	Chance (1980)
		15,170	Tokura *et al.* (1984)
ETCD[c]		15,750	Chance *et al.* (1977)
		130°C: $\sim$18,500	

[a] Does not coincide with the absorption peak! The shift is of the order of 100 cm^{-1}.

[b] Slightly redshifted at lower T.

[c] Does not polymerize completely. The remaining $\sim$15% monomer may be extracted. Most published data are on such extracted material.

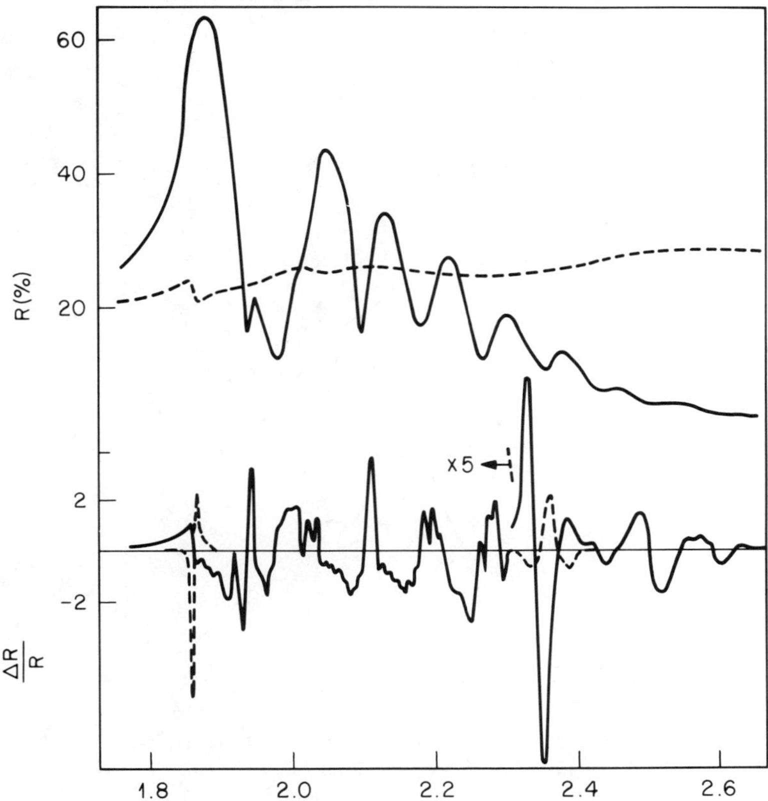

Fig. 8. Optical spectra of DCH. The upper curves show the reflection spectrum at 8 K in percent reflectance, and the lower ones the electroreflectance at 2 K, in units of 10^{-3}. Solid lines, light polarized parallel to the chains; dotted lines, light polarized perpendicular to them. [Adapted from Sebastian and Weiser (1981a), with permission.]

Table III lists two cases (ETCD and IPUDO) where the two types of absorptions occur in the same crystal, by varying the temperature. Figure 9 shows the case of IPUDO; spectra at 25 and 135°C are compared to room-temperature spectra of DCH and TCDU-I (Eckhardt *et al.*, 1979). In the case of TCDU, two monomer crystal modifications are known, which upon reaction yield different polymers, but phase I of the polymer can be more or less completely transformed into phase II by cooling, hydrostatic pressure (Iqbal *et al.*, 1977), or "strain" (Tokura *et al.*, 1984).

This at least shows that the electronic properties of a given polydiacetylene can vary; they are not uniquely determined by its side-group molecular structure, but they result from interactions, not yet understood, with their environment as a whole. Examination of the changes in optical absorption of

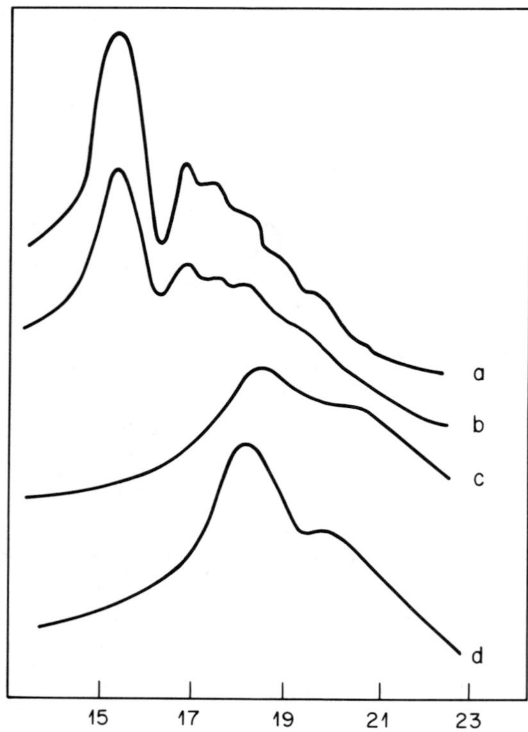

Fig. 9. The two types of polydiacetylene reflectance spectra. (a) DCH, (b) IPUDO at 300 K, (c) IPUDO above 400 K, and (d) TCDU-I. [Adapted from Eckhardt *et al.* (1979), with permission.]

the polymer chains dispersed in the monomer matrix—which will not be discussed here—would lead us to the same conclusion (see, for example, Bloor and Hubble, 1978). These interactions are further discussed in Section III.C.2.

The color change of a monomer crystal upon polymerization is spectacular, and a crystal containing $\sim 1\%$ polymer is already highly absorbing, so it is often said that the transition responsible for the visible absorption spectrum of polydiacetylenes is a very intense one. In fact, its oscillator strength has not been measured very accurately. Several quantitative observations suggest that it is of the order of 1. First, although ε_1 becomes negative in PTS near $16,000$ cm^{-1}, the reflection spectrum shows no obvious stop band at any temperature, contrary to many aromatic hydrocarbon or dye crystals for transitions with $f > 1$ (for a review, see Philpott, 1973, 1980), but this may be due to the dilution of the chains into a nonresonant medium made of side groups, so that there is only one repeat unit for >500 Å^3, whereas in anthracene, for instance, there is one moelcule per 240 Å^3. The (100) natural face of PTS (which contains the chain direction) has been shown to support surface polariton modes at room temperature, albeit in a small energy range only (Brillante *et al.*, 1978; Philpott, 1980).

A very large transition dipole would correspond to large dispersion of the one-dimensional exciton band, whereas the experimental spectrum shows regular progressions of totally symmetric modes rather reminiscent of a weak coupling case, including some very low energy modes, below 100 cm^{-1} (Sebastian and Weiser, 1981b). Thus, the available evidence is not fully consistent yet.

There is no obvious further transition in the visible, and the near-UV transitions can all be assigned to side groups; this has been studied for instance on DCH by Hood *et al.* (1978). Tokura *et al.* (1980) studied the reflectivity of polymer and monomer PTS single crystals up to 20 eV (Fig. 10). The spectrum

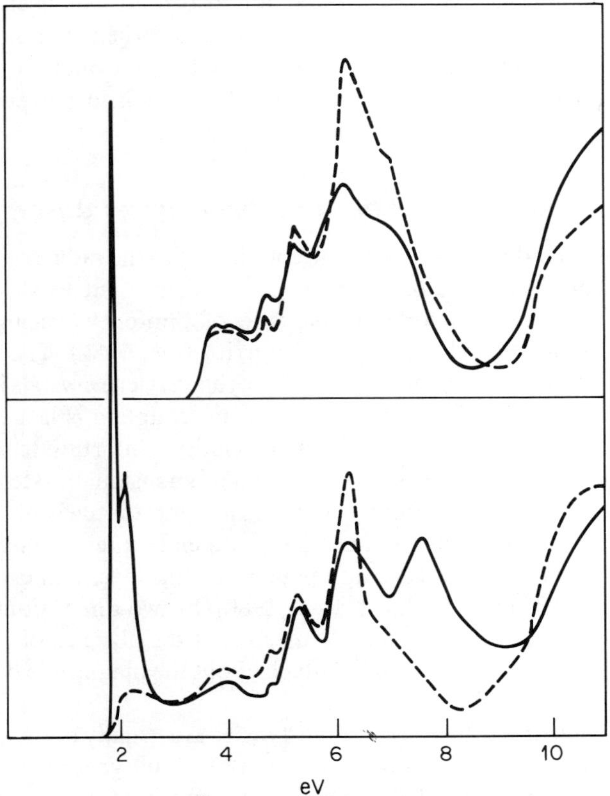

Fig. 10. Reflectance spectra recorded by Tokura *et al.* (1980). Upper curves are for monomer PTS, lower curves for the corresponding polymer. Light polarization parallel (solid lines) and perpendicular (dotted lines) to the crystal *b* axis, which is, in the polymer, parallel to the chain direction. [From Tokura *et al.* (1980), with permission.]

contains the already discussed transition from 2 to 3 eV, and several side-group absorptions from 4 to 7 eV, which are not affected by polymerization. A diacetylene monomer transition near 7 eV disappears on polymerization and is replaced by a new peak at 7.6 eV. The assignment by Tokura *et al.* of this peak to a Wannier exciton of the polymer chain is unreasonable, since the bandgap is around 2.5 eV (see Section III.D). Above 8 eV the spectrum is broad and structureless, so the optical spectrum, except below 3 eV, contains little readily useful information about the polymer electronic properties. We shall return to two-photon absorption and to electroreflectance in Section III.D.5.

No other electronic transition has been found below the 2-eV absorption. It tails toward the red in some polymers, in what appears to be associated with absorption by surface states and other defects (Reimer and Bässler, 1975; Weiser and Sebastian, 1985). In the IR, no absorption has been found that could not be assigned to molecular vibrations and their combinations and harmonics (Eichele *et al.*, 1980). This is not enough to rule out the existence of a low-lying triplet, or a totally symmetric singlet state as in polyenes (Hudson *et al.*, 1982).

2. Crystal Environment Effects on the Polymer Absorption

In a strictly one-dimensional crystal, coupling with the radiation field differs from that usual in three dimensions in a way equivalent to shortening the radiative lifetime by a factor λ/a, the ratio of photon wavelength to one-dimensional unit cell length, or $\sim 10^3$ (Orrit *et al.*, 1982). The absence of polymer luminescence from good-quality polydiacetylene crystals implies that they do not show this effect, so that interchain coupling is large enough to destroy the required degree of one-dimensionality. Unfortunately, the corresponding theory of very weakly coupled one-dimensional systems in interaction with a radiation field is not available. Similar radiatively unstable states are observed in quasi-two-dimensional systems, the surface layer of anthracene crystals, where the coupling between this layer and crystal bulk is of the order of 10 cm^{-1} (Philpott, 1973, 1980). In two dimensions, however, the factor of interest is $(\lambda/a)^2$, not (λ/a), so that the absence of the effect in polydiacetylene crystals does not tell much about the magnitude of interchain coupling.

The electronic properties of polydiacetylenes are usually discussed theoretically using one-dimensional models and simplified side groups, often replaced by H or CH_3 (see Section III.D). This neglects several important factors. The chemical nature of side groups may of course influence the chain geometry, dynamical properties, and electron transition energies; this was discussed already above. In addition, several three-dimensional processes exist, which not only introduce some three-dimensional coupling, but also modify the "one-dimensional" electronic properties.

Let us consider first interchain coupling processes. Bulky side groups separate the chains, but packing may be such that quite close interchain approach may occur in one direction, creating preferential two-dimensional coupling between chains. Interchain closest approach may be as large as 9 Å in DNP or DCH, or 7.5 Å in PTS; in these cases, there is no real preferential direction for coupling, and the closest chains are not translationally equivalent. When side groups are longer and are extended in the crystal, as in TCDU-1, 4pBrPU, or PTS-12, translationally equivalent chains can be as close as 5.7 Å (see references in Table II). A distance of 4.8 Å has even been inferred (on the basis of unit cell dimensions) for the high-temperature phase of ETCD (Chance *et al.*, 1977). For comparison, the smallest distance between molecular centers in crystals of small aromatic molecules like naphthalene is of the order of 5.2–5.5 Å, and the smallest interchain distance in polyacetylene is 4.2 Å (Fincher *et al.*, 1982). Thus, interchain distances in polydiacetylene crystals are comparatively very large, and corresponding interactions are expected to be small.

Coupling between chains will generate a nonzero bandwidth in directions perpendicular to the chains. In an exciton picture, coupling between translationally nonequivalent chains will generate a splitting of optical absorption lines, with different polarizations for the different components, whose magnitude is dependent on interchain interaction matrix elements (Davydov, 1962, 1971). Such splitting has been searched for. In all polydiacetylenes, the reflection spectrum polarized perpendicular to the chain direction is almost structureless, but weak features are seen (Fig. 8). In PTS at 300 K, Müller and Eckhardt (1978) found a weak a polarized structure displaced from the b polarized peak; Kramers–Kronig analysis yielded a splitting < 100 cm^{-1} with a dichroic ratio $\varepsilon_a/\varepsilon_b \approx 2 \times 10^{-2}$. The accuracy is limited, but the existence of this splitting was confirmed on low-T spectra (at least for the higher-lying transition), and is seen on electroreflectance spectra as well (Sebastian, 1980). Nonequivalent chains are still parallel, and the angle of the transition dipoles with their common direction is very small: a value of 4 degrees has been experimentally inferred in PTS (Bloor and Preston, 1977) and TCDU (Müller *et al.*, 1977), and calculated in the Hückel model (Chance *et al.*, 1980a). This may explain the small value of the splitting, which is therefore not a definite proof of a small interchain interaction.

The influence of side groups on energy levels and transition energies, not so much via direct coupling to the π electrons than via polarization effects (solvent shift, in chemical terms) is probably more important than interchain resonance interactions. A dramatic experimental illustration is provided by the low-T behavior of PTS. In this material, a second-order phase transition occurs at ~ 190 K with a doubling of the unit cell along a (Enkelmann, 1977; Bloor *et al.*, 1979). The low-temperature cell contains two groups of two chains, not related to each other by any symmetry element. In fact, the interchain distances do not change and the geometries of the chains remain identical, but the corresponding side-groups orientations differ slightly (Enkelmann, 1977). All optical absorption peaks are split by an equal amount, 300 cm^{-1} at 4 K (Bloor and Preston, 1977); this can be interpreted as a site splitting: individual chain transition energies may remain equal, but the interactions of each chain with the surrounding medium become different in the ground and in the excited states (Schott *et al.*, 1978).

This observation shows that the observed transition energies are sensitive, at least to the 0.1-eV level, to seemingly slight differences in their environment—for instance, in the case of PTS, to orientation of neighboring dipoles (those of the toluene sulfonate groups). Thus three-dimensional perturbations to several properties of this apparently one-dimensional system are far from negligible. After we have discussed, in the next section, the nature of the excited state responsible for this absorption, we shall return briefly to a discussion of the effect of crystal environment on this excited state.

3. The Effect of Strain

In a real crystal, the polymer backbone may be strained. Mechanical response of polydiacetylene fibers has been studied for years, since one potentially interesting property of these polymers is their high modulus. A recent review of these properties is given by Young (1985). Some polydiacetylenes form naturally single-crystal fibers suitable for optical study under well-defined strain, obtained by tension parallel to the fiber and chain direction. Raman frequencies of the chain vibrations can be selectively studied under resonance conditions, optically exciting only a small fraction of the fiber length. This allows us to check the homogeneity of the strain and to measure its influence on the force constants. For instance, a linear variation of the frequencies up to 1.7% strain along the chain direction has been found on DCH—a favorable case (Batchelder and Bloor, 1979)—so the geometry and electronic properties vary regularly. Similarly, the absorption of PTS shows a gradual blueshift upon increasing uniaxial stress along the chain direction (Batchelder and Bloor, 1978) of 37 meV per percent elastic strain.

Elastic deformation of the crystal does not result solely in chain extension, which affects intrachain properties: several couplings are allowed by the low symmetry of polydiacetylene crystals. The corresponding elastic constants are not negligible (Leyrer *et al.*, 1978; Rehwald *et al.*, 1983), so strain-produced environment changes may affect the electronic properties, as discussed in Section III.C.2. The latter effects have not be considered up to now; they may be small.

On the other hand, when hydrostatic pressure is applied, chain lengths are largely unaffected: the compressibility of PTS along the chain direction is almost zero (Lochner *et al.*, 1980), and this is taken to be a general property of polydiacetylenes. The first-order effect of hydrostatic pressure is then to bring the chains closer together and compress the side groups, so changes in the environmental effects discussed in Section III.C.2 are dominant (see Section III.D).

4. Resonance Raman Scattering

This is a very useful method for the study of conjugated molecules with complicated side groups like polydiacetylenes, since only totally symmetric vibrations coupled to the backbone electronic transition are resonantly enhanced, allowing us to study electronic and mechanical properties of the backbone (Batchelder and Bloor, 1982). Although the method is an important and powerful one and yielded important informations on polydiacetylenes, it will not be further discussed here, since it has been extensively and adequately covered in a recent review by Batchelder and Bloor (1984).

D. The Nature of the Visible Optical Transition of Polydiacetylene Crystals

1. Exciton or Band-to-Band Transition?

Since the earliest spectroscopic studies on polydiacetylenes, the nature of the first optically allowed transition has been debated. Although an excitonic origin was proposed very early (Bloor *et al.*, 1974), it was initially mostly discussed in terms of band-to-band transition, neglecting electron correlation (Wilson, 1975), as in calculations of linear and nonlinear optical properties (Cojan *et al.*, 1977; Agrawal *et al.*, 1978), for instance. Today, it is fairly generally accepted that the lowest excited state is excitonic. This is partly due to recent acceptance of the important role of electron correlations, and to their incorporation into more and more sophisticated calculations, and partly to experimental evidence, an important piece being the absence of photoconductivity at gap energy, at least in some polydiacetylenes. However, it is not generally accepted that electron correlation plays a dominant role in the somewhat similar polymer polyacetylene, where the exciton concept is only slowly getting consideration, if at all. Also it may be wise to beware of fashion, so a short consideration of the experimental evidence may be in order.

The reflection spectra and the absorption ones obtained by KK inversion begin with a strong and narrow band—we neglect here the complications, peculiar to PTS, of site splitting—followed toward higher energies by a series of weaker bands, particularly rich in PTS-like spectra. The energy difference between the first band and any of the others is close to a vibrational quantum of the polymer chain as observed in resonance Raman scattering (Batchelder and Bloor, 1984). Such spectra occur naturally in moderately strong exciton transitions in the weak coupling case, where the narrowness of the first, purely electronic, transition and of the others is due to momentum conservation, and vibronic transitions are separated from the electronic one by excited-state vibrational quanta. That the purely electronic transition is the more intense, and that the observed energy difference are near the ground-state vibrational energies mean that the excited-state geometry is not very different from the ground state one (large Franck–Condon factor). In the band picture, the steep rise corresponds to the singularity of the density of state at band edge in this quasi-one-dimensional system (van Hove singularity), in conjunction with a direct gap (Cojan *et al.*, 1977). The shape of the spectrum above the first maximum depends on parameters like relaxation times, delocalization, etc., and on coupling strengths with molecular vibrations that have not been calculated. Pending such calculations, the optical absorption itself cannot be used as a proof for or against any of the models.

It is fair to say that, were the band picture correct, exciton effects would be completely absent, since there is no absorption, however weak, below the threshold, that cannot be ascribed to defects. On the other hand, the exciton picture implies that band transitions are not clearly visible in the spectrum; this and the large intensity of the exciton transition implies that the exciton binding energy is at least several tenths of an electron-volt, as in the well known polyacene crystals (Pope and Swenberg, 1982).

2. Evidence from Photoconductivity Experiments

Although conductivity in itself is out of the scope of this review, it is important to mention the photoconductivity action spectra, as they provide information on the nature of the lowest excited state. In some cases, the photoconductivity threshold is well above the absorption threshold. Figure 11 shows the case of DCH (Lochner *et al.*, 1981, see also Yee and Chance, 1978), where the (ill-defined) conductivity threshold is around 18,000 cm^{-1}, 2700 cm^{-1} above the optical threshold. Similar results were obtained on polydiacetylene multilayers (Lochner *et al.*, 1978). In PTS (Siddiqui, 1980; see also Chance and Baughman, 1976) or TCDU (Lochner *et al.*, 1978), on the contrary, the photocurrent threshold is well below the absorption threshold near 6500 cm^{-1} in PTS. Such a low energy threshold may be associated with defect photoionization, or with photoionization from the electrodes (Pope and Swenberg, 1982), and is certainly extrinsic. The steep rise at energies above the absorption threshold follows approximately a $v^{1/2}$ law (Lochner *et al.*, 1978) that extrapolates to zero yield around 21,000 cm^{-1} in TCDU and 17,000 cm^{-1}

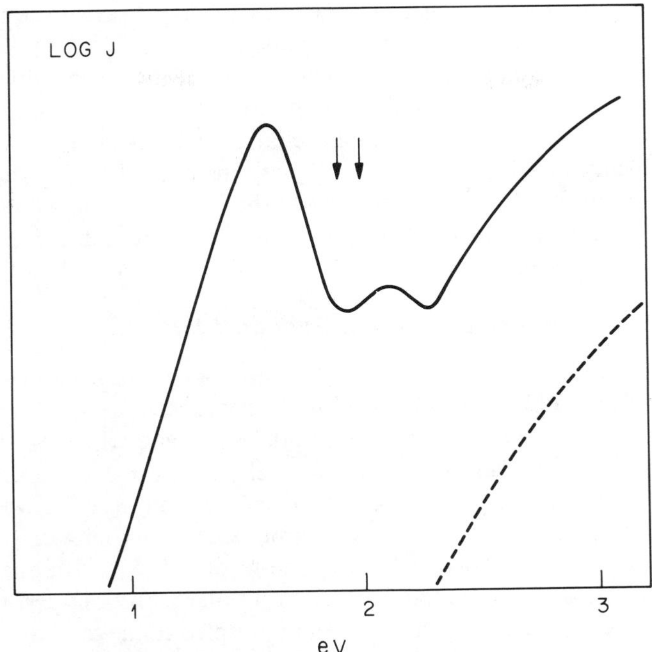

Fig. 11.   Photoconductivity action spectra for PTS (Siddiqui, 1980), solid line, and DCH (Yee and Chance, 1978), dotted line. The arrow at right corresponds to the absorption threshold for PTS, the arrow at left to that of DCH. Currents are on a logarithmic scale.

in PTS. If this corresponds to an intrinsic process, the corresponding threshold is above the absorption one, although barely so in PTS.

These observations are not easily accounted for by a one-electron model for the optical absorption. One might like to invoke surface recombination decreasing the yield at maximum absorption. This, however, would yield a relation between yield and absorption coefficient above and below the peak absorption, which does not seem to be observed.

It would then seem natural to consider this as proof of the exciton model for the absorption. The exciton would then not ionize during its lifetime, which is short (in the picosecond range), since it does not fluoresce to any measurable extent (Enkelmann *et al.*, 1977). This short lifetime explains the absence of exciton-sensitized photoconduction so common in organic solids (Pope and Swenberg, 1982, and reference therein). Note, however, that the result of the experiments is not clear, since the inferred binding energy in PTS is small.

Thus, in the exciton picture, which would appear to be supported by these experiments, polydiacetylenes have around 20,000 cm^{-1} a state precursor to electron–hole pair generation. It does not, however, necessarily coincide with a band-to-band transition, since one may also assume the existence of another autoionizing state, the gap then being smaller.

If we stop the discussion here to conclude that the lowest transition is to an exciton state, this leads to two further questions: what is the nature of this exciton, and where is the one-particle excitation gap? We shall now consider these problems, and leave aside other questions related to conductivity, in particular electrical charge transport, since they are not directly in the theme of this book. Discussions of these problems may be found, for instance, in Bässler (1985) or Wilson (1985).

3. A Theoretical Model for the Exciton State

Philpott (1977) was the first to study theoretically an exciton model for polydiacetylenes, following a semiempirical approach already developed by Pugh (1973) and others. He found that both singlet and triplet excitons have large charge transfer (CT) character, with a delocalization over approximately two monomer units. This was developed by Yarkony (1978), who indeed found, treating electron correlations with the ZDO approximation, that the exciton is the lowest level, with a binding energy of ~ 1 eV and about the same CT character as found by Philpott (1977). Molecular vibrations perturbing the CT mixing are expected to be most strongly coupled to the transitions: $C{=}C$ stretch and torsions.

The most recent theoretical treatment is that of Suhai (1984), who calculated the visible absorption spectrum of charge-transfer excitons in single polydiacetylene chains for the geometry of PTS and a geometry similar to that published for TCDU, as a model of a butatrienic structure, by a nonempirical method with particular attention given to electron-correlation effects. The details of the calculation will not be reported here, nor the various approximations be discussed, but the main results will be given to show where such calculations fit or not with what is experimentally known. The lowest optically accessible excited state is a somewhat delocalized (charge transfer) exciton: excitation energies converge only if its radius is > 25 Å, so it is very far from the Frenkel limit (see Davydov, 1971). This is a larger delocalization than found in previous semiempirical calculations (Philpott, 1977; Yarkony, 1978), and also larger than assumed to fit electroreflectance spectra (Sebastian, 1980). For PTS, the transition energy is calculated to be ~ 2.2 eV; it cannot be simply compared with the experimental value of 2.0 eV, since the latter includes dipolar interactions with the side groups (see Section III.D.4), which have not been calculated, but the order of magnitude is certainly correct. The exciton band is > 2 eV wide, which is very large and raises question about the applicability of weak-coupling vibronic theory. A narrower triplet exciton band is predicted around 0.8 eV. The singlet exciton binding energy is ~ 0.4 eV, which is physically reasonable, otherwise the exciton oscillator strength would be smaller than observed; the ionization potential is 5.7 eV. Again, considering that interaction with the side groups has not been taken into account, the agreement with experiment [$\sim$0.4 eV exciton binding energy from electroreflectance; $\sim$5.4 eV ionization potential (IP) (Arnold, 1982; Murashov *et al.*, 1982)] is satisfactory. This lends credence to the results of Suhai's calculation for the other geometry.

This other geometry is similar to, but unfortunately not identical with, the minimum energy geometry calculated by Karpfen (1980) for the butatrienic ground state. Suhai's conclusions are striking: in this geometry, the exciton transition energy is $\sim$0.4 eV *below* that of the enyne structure. This is comparable to the difference in excitation energy between butatrienic diradicals and enyne asymmetric carbenes of the same length, as observed in low-temperature photopolymerization (Sixl, 1984). The exciton binding energy is comparable, and the ionization potential is about 5.0 eV. The difference in IPs for the two geometries agrees with the result of the valence effective Hamiltonian calculation of Bredas *et al.* (1981). Therefore, the calculation does not reproduce at all the properties of TCDU or of any "butatrienic" polydiacetylenes, suggesting that the origin of their blue-shifted spectra is different.

The overall agreement between this theory and experiment is therefore fairly good, and allows us to conclude that the state responsible for the lowest optical absorption of polydiacetylenes is an exciton with an extension of 15–30 Å. The large exciton bandwidth would need experimental confirmation. Note that Batchelder and Bloor (1984) could well account for resonance Raman excitation spectra neglecting exciton dispersion. It would now be interesting to consider nonlinear optical responses in the same theoretical framework.

4. Environmental Effects

It would also be useful to consider explicitly the influence of the crystalline environment of polydiacetylene chains on their electronic excitations. In molecular crystals, where the Frenkel limit applies, exciton energies are often written (see, for instance, Philpott, 1973) as

$$E(\mathbf{k}) = E_0 + D + I(\mathbf{k}) \tag{7}$$

where E_0 is the isolated molecule transition energy, D the so-called gas-to-crystal shift, and I the matrix of exciton transfer interaction between molecules, which is usually broken in two parts corresponding to interaction between translationnally equivalent and nonequivalent, molecules, respectively, the latter part giving at $\mathbf{k} = 0$ the so-called Davydov splitting (Davydov, 1962, 1971).

Polydiacetylene crystals, which we take as built from one-dimensional polymeric molecules, differ from conventional molecular crystals in many ways. Since the "molecules" are in fact one-dimensional crystals, E_0 is replaced by a dispersion relation $E_0(k_{||})$ with a bandwidth that may be comparable to the transition energy. The energy shift and resonance interaction terms—the latter being now $I(\mathbf{k}_\perp)$—may vary with $k_{||}$, so the shape of the one-dimensional band may be affected by three-dimensional interactions. In addition, E_0 is well defined for a rigid molecule only, whereas here it depends on the chain geometry and may be quite sensitive to it (this would be calculable). The energy shift may be large and widely variable from one polydiacetylene to another. One may consider the chains as "diluted" in a polarizable medium provided by the side-groups. Let us then boldly consider a chain as if in a dilute solution in a side-group solvent, and use a formula like the one proposed by Longuet–Higgins and Pople (1957) for the solvent shift W (which certainly provides the larger part of D):

$$W = \frac{\alpha_s n}{6R^6}\left|M^2 + \frac{E\alpha_c}{4}\right| \tag{8}$$

where a chain of polarizability α_c with an electronic transition of dipole moment $\mathbf{M}$ and energy E is surrounded at a distance R by n side groups of polarizability α_s. Large chain polarizabilities have been quoted, a value 10^{-21} cm^3 inferred from electroreflectance measurements (Weiser and Sebastian, 1985), so a shift larger than 1000 cm^{-1} is easily obtained. But in addition, many side groups, as in PTS, DNP or DCH, bear permanent dipole moments of the order of 1D, and W may be even larger.

As discussed by Batchelder (1985), following Rice and Jortner (1965), an estimate of W (neglecting the complications due to dipole reorientation) can be deduced from the pressure dependence of the transition energy (Cottle *et al.*, 1978; Jankowiak *et al.*, 1978; Lacey *et al.*, 1984). One finds $W \approx 2500$ cm^{-1} for PTS, or about 15% of the transition energy. A better understanding of the influence of the crystalline environment on the chain electronic properties is certainly needed. It might contribute to explaining (besides via "strain") the variation of polydiacetylene transition energies.

5. *Other Excitations in the Visible and Near-UV*

Various spectroscopic techniques have shown the existence of other electronic excitations a few electron-volts above ground state, not accessible in one-photon absorption processes. The techniques were electroreflectance, two-photon absorption, and electron energy-loss spectroscopy, showing band-to-band transition, A_g state, and valence plasmon excitation.

Sebastian and Weiser (1979, 1981a,b) studied the electroreflectance spectrum of several polydiacetylenes—PTS, DCH, pFBS—at room temperature and around 10 K. Such spectra show relatively rich structures in the exciton absorption region, which the authors used to infer an order of magnitude of the exciton delocalization (~ 10 Å), and at low temperature a more intense one at somewhat higher energy, which they interpreted as due to the onset of band to band transition. In this energy range, the reflexion spectrum is smooth and $Im\varepsilon$ is small, so the corresponding oscillator strength must be small. Applying standard interpretation of electroreflectance spectra of semiconductors yielded a small effective mass value $m^* \approx 0.05$, in accordance with the wide calculated bands (Sebastian and Weiser, 1981a). The experimental splitting between the exciton and band-to-band signals was on the order of 0.5 eV.

The assignment of the large electroreflectance signal to a band-to-band transition has been recently questioned by Tokura *et al.*, (1984), who claim—without, however, a detailed reinterpretation of the data—that it is actually due to the ^1A_g singlet exciton, the applied field lifting the forbiddenness of the $^1A_g \rightarrow {}^1A_g$ transition.

Such a state certainly exists. In long polyenes, it is the lowest-lying one, and anomalous fluorescence and strongly allowed two-photon absorption are connected to it (Hudson *et al.*, 1982). In the case of polydiacetylene crystals, no two-photon state has been found below the one-photon threshold, but a few two-photon absorption experiments have been performed. Reimer and Bässler (1978) found $\beta \approx 5 \times 10^{-48}$ cm^4 sec/photon per repeat unit at a two-photon energy of 18,880 cm^{-1} in PTS (~ 2600 cm^{-1} above the one-photon threshold). Earlier results of Lequime and Hermann (1977) on PTS and TCDU, which these authors interpreted in terms of photogeneration of real transient states with picosecond lifetime, were reinterpreted by Chance *et al.* (1980a) in terms of two-photon absorption, in agreement with the latter authors' work on polydiacetylene solutions, yielding $\beta \approx 7 \times 10^{-47}$, a very large value, near a two-photon energy of 23,500 cm^{-1} in PTS.

Hückel-type calculations indeed predict a strongly two-photon allowed transition to such a state in the correct energy range, about 0.5 eV above the lowest exciton (Chance *et al.*, 1980a). This agreement might seem surprising, bearing in mind that Hückel calculations grossly overestimate the energy of the corresponding state in polyenes (see, for instance, Schulten *et al.*, 1976). A calculation including correlation effects on simple enynes, divinylacetylene and 1,5,9-decatriene-3,7-diyne, shows however (Dinur and Karplus, 1982) that correlation corrections are smaller in enynes than in polyenes and that the ^1A_g state is not the lowest-lying excited state. This conclusion would seem to extend to longer enynes and polydiacetylenes as well, and is compatible with the ^1A_g state being ~ 0.5 eV above the lowest exciton in polydiacetylenes. This state would play a very important role in nonlinear optical properties (Chance *et al.*, 1980a).

Dinur and Karplus (1982) also found that certain $\sigma \to \sigma^*$ transitions have energies comparable to the $\pi \to \pi^*$ ones, so that $\pi\sigma$ interactions should be carefully taken into account in the case of polydiacetylenes. This finding is a difficulty for Hückel-type calculations.

Finally, collective excitations of π electrons are known to occur in crystals of conjugated molecules and are observed in electron energy-loss spectroscopy (EELS). A peculiar feature of polydiacetylene crystals is the low density of polymer-chain π electrons, which are "diluted" by the side groups in the three-dimensional crystal. The corresponding plasmon energy is therefore low, less than 4 eV when evaluated using the theoretical expressions of Horie (1959) or Egri (1985). Features at this low energy are indeed found in EELS (Ritsko *et al.*, 1982; Rei Vilar, 1985). These collective excitations should be taken into account in the evaluation of the one-particle excitation spectrum. Their influence on nonlinear optical properties of polydiacetylenes has never been considered.

E. Electronic Properties of Less Well Ordered Polydiacetylenes

Several less well ordered states of polydiacetylenes are known: Langmuir–Blodgett monolayers and multilayers, solutions, gels, fibrillar precipitates, and liquid-like states reminiscent of liquid crystals (already discussed).

1. Single Chains (Yellow Solutions)

The only cases where single polymer chains, not interacting with one another, are present are in true dilute solutions.

Solubility was observed and used for specific viscosity measurements very early in the study of polydiacetylenes (Wegner, 1971b). Several families of polydiacetylenes are now known to be soluble in suitable solvents. The first extensively studied series contains long aliphatic side groups bearing amide functionalities, so that H bonds can be formed between neighboring side groups of the same chain: the so-called 3-BCMU and 4-BCMU (Patel, 1978a,b; for recent discussions, see Chance *et al.*, 1985; Lim *et al.*, 1985 and references therein). Several other families, not all containing amide groups, are now known (Wenz *et al.*, 1984; Plachetta *et al.*, 1982; Plachetta and Schulz, 1982, etc...). All these polymers form in good solvents true solutions with very similar visible absorption spectra: a broad, structureless peak with a threshold near 550 nm and a peak around 470 nm (Fig. 12), suggesting that the nature of side groups and their interactions between themselves or with solvent molecules have limited influence on the electronic structure of the chain.

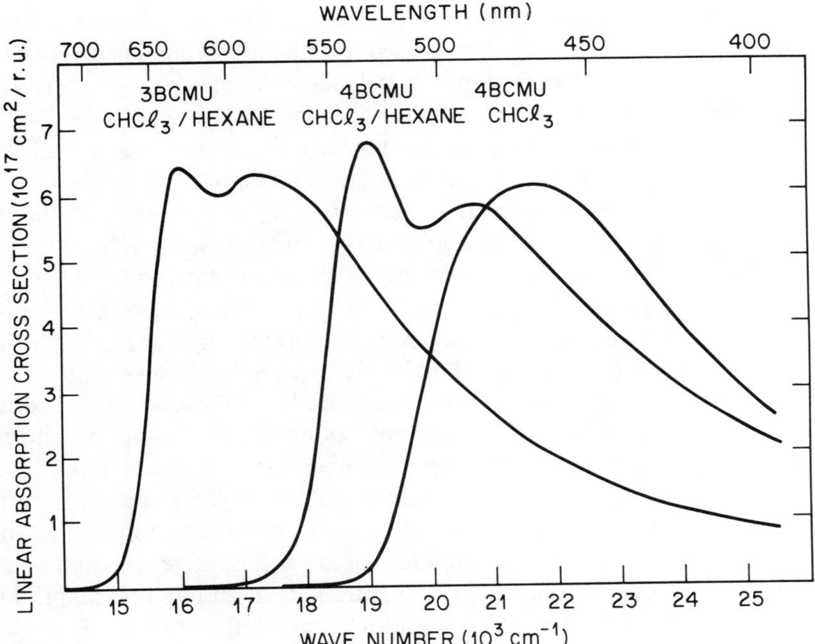

Fig. 12. Absorption spectra of the yellow, red, and blue "solutions" of 4-BCMU and 3-BCMU.

Two-photon absorption in yellow solutions was studied for 3- and 4-BCMU (Chance *et al.*, 1980a) and PTS-12 (Kajzar and Messier, 1985). The corresponding A_g state was placed near 30,000 cm^{-1} in the former case, and below 28,200 cm^{-1} in the latter. These values, however, are not accurate, and the whole two-photon absorption spectrum, which is certainly inhomogeneously broadened, is not known, so we cannot say if it is essentially the same for all yellow solutions, as is the one-photon absorption. ^{13}C-NMR indicates an enyne structure (Babbitt and Patel, 1981), and neutron small-angle scattering shows that they are in the trans isomer form (Rawiso *et al.*, 1986). The chains are coiled, but not very flexible. They are best represented by a worm-like chain model with a persistence length of about 160 Å at room temperature; the same value is obtained for polymers like 3-BCMU with H-bonding side groups (Rawiso *et al.*, 1986) and for PTS-12, which cannot form H bonds (Wenz *et al.*, 1984), suggesting that the conjugated chain conformation is not very sensitive to the nature and properties of its side groups.

The absence of structure in the absorption band has been shown by resonance Raman (RR) scattering to be related to inhomogeneous broadening (Shand *et al.*, 1982). The blue shift, compared to crystal absorption, is related to a less conjugated structure in the solution. The only model developed up to now treats the chain as a collection of uncoupled boxes of different lengths; the electron states of each box may then be calculated for electrons in a box with periodic potential (Exarhos *et al.*, 1976) or by standard Hückel theory (Chance *et al.*, 1980a), for instance. An absorption near 470 nm corresponds to a box containing seven monomer repeat units, and the threshold absorption to ~ 15 units. The length of a box is then referred to as a conjugation length. Although this model has been a useful guide in understanding the RR results, it fails to explain the complete absence of structure of the absorption spectrum, unless unphysically large linewidths are assumed (Shand *et al.*, 1982). Certainly interbox coupling should exist. But more important, the concept of a box containing an integral number of perfectly ordered monomer units is not compatible with either the worm-like chain model or the neutron-scattering results, which indicate that rigid segments along the chain must be shorter than about four units (corresponding to an absorption near 400 nm).

A better model is not out of reach: a polydiacetylene chain is a one-dimensional semiconductor along which potential fluctuates in space and time for several reasons: relative orientation of neighboring repeat units, side groups, and solvent molecules. These fluctuations, even of small amplitude, localize the electronic states. A conjugation length is then the extension of a wave function: it is no longer restricted to integral multiples of a repeat unit length. No strong potential barrier is needed (such barriers, associated to localized defects, may exist as well). In such a model, the conformational informations obtained by scattering experiments would be used to calculate one-dimensional electronic states in the presence of disorder, the more important being probably the nondiagonal one. Conjugation lengths in the latter sense may be different for different electronic states: those of the ^1A_g two-photon accessible state associated to the large third-order susceptibilities may be different from those of the ^1B_u state responsible for one-photon absorption. This conjecture is open to experimental and theoretical testing.

Pending the development of such a model, the earlier one is still a useful zero-order approximation, at least for qualitative discussion.

2. The Color Transitions

Starting from a yellow solution, a change in solvent quality in the case of BCMU polymers (Chance *et al.*, 1985; Lim *et al.*, 1985) or pH or ionic strength in the case of polyelectrolytes (Bhattacharjee *et al.*, 1980) produces changes in conformation (as discussed already) and in electronic properties of the polydiacetylenes: absorption spectrum and resonance Raman frequencies change (see, for instance, Shand *et al.*, 1982). The resultant phase may be fluid, or a gel, or a fibrous precipitate. The most common absorption spectrum peaks around 540 nm, 4-BCMU being a typical example, and the medium is red. In some cases, 3-BCMU being an example, the shift is larger, the absorption peaks around 630 nm, and the medium is blue (Fig. 12). But the same polymer in different conditions may yield one of the two colors or the other.

These changes are reversible: upon increase in solvent quality, the yellow solution is recovered. In other imperfectly organized phases, as in LB films of certain polydiacetylenes, *irreversible* changes are observed [see Tieke *et al.* (1979, 1981) and Lieser *et al.* (1980) for an example of a "blue-to-red" change] that are more similar to (sometimes reversible) color changes occurring at phase transition in polydiacetylene crystals.

The absorption spectra of the red and blue phases are strongly reminiscent of those observed in single crystals. Indeed, there is reason to believe that in all cases, except perhaps the most dilute solutions (Lim *et al.*, 1985), polymer chains are aggregated; some kind of local order of elongated chains must exist, extending along the chains over long enough distances. The irreversible changes observed in solid phases may be related to structural changes produced by removal of solvent or unreacted monomer, or simply because the structure directly obtained by polymerization is metastable, or else due to polymer degradation. Unfortunately, the rapidly growing number of experimental observations does not yet fit into a consistent picture.

IV. CONCLUSION

In this review, we did not try to present a complete, nor even a balanced, description of polydiacetylenes preparation and properties. Only topics of relevance to the main theme of this book were discussed, and among them emphasis was put on some problems that, while still controversial, are advanced enough to warrant a careful discussion. No doubt, in so doing, the authors' prejudices play a role. We hope that the interested reader will be able to build his or her own prejudices by using the original references cited, so that this review will have been of some use to some people, after all.

ACKNOWLEDGMENT

One of us (M.S.) gratefully acknowledges the very kind hospitality of Professor M. Schwoerer at the University of Bayreuth, where part of this review was prepared.

REFERENCES

Agrawal, G. P., Cojan, C., and Flytzanis, C. (1978). *Phys. Rev.* **B17,** 776.

Aimé, J. P. (1983). Thesis, Université Paris VII (unpublished).

Aimé, J. P., Lefebvre, J., Bertault, M., Schott, M., and Williams, J. O. (1982). *J. Phys. Orsay, Fr.* **43,** 307.

Albouy, P. A. (1982). Thesis (3ème cycle), University of Orsay, France, unpublished.

Almenningen, A., Bastiansen, O., and Traetteberg, M. (1961). *Acta Chem. Scand.* **15,** 1557.

Almenningen, A., Gogstad, E., Hagen, K., Schei, H., Stølevik, R., Tringstad, Ø., and Traetteberg, M. (1984). *J. Mol. Struct.* **116,** 131.

Apgar, P. A., and Yee, K. C. (1978). *Acta Crystallogr., Sect. B* **B34,** 957.

Arnold, S. (1982). *J. Chem. Phys.* **76,** 3842.

Babbitt, G. E., and Patel, G. N. (1981). *Macromolecules* **14,** 554.

Bader, H., Dorn, K., Hupfer, B., and Ringsdorf, H. (1985). *Adv. Polym. Sci.* **64,** 1.

Bässler, H. (1984). *Adv. Polym. Sci.* **63,** 1.

Bässler, H. (1985). *In* "Polydiacetylenes" (D. Bloor and R. R. Chance, eds.), pp. 135–154. Martinus Nijhoff, Dordrecht, Netherlands.

Batchelder, D. N. (1985). *In* "Polydiacetylenes" (D. Bloor and R. R. Chance, eds.), pp. 187–212. Martinus Nijhoff, Dordrecht, Netherlands.

Batchelder, D. N., and Bloor, D. (1978). *J. Phys. C* **11,** L629.

Batchelder, D. N., and Bloor, D. (1979). *J. Polym. Sci., Polym. Phys. Ed.* **17,** 569.

Batchelder, D. N., and Bloor, D. (1982). *J. Phys. C* **15,** 3005.

Batchelder, D. N., and Bloor, D. (1984). *Adv. Infrared Raman Spectrosc* **11,** 133–209.

Baughman, R. H., and Chance, R. R. (1978). *Ann. N.Y. Acad. Sci.* **313,** 705.

Baughman, R. H., and Yee, K. C. (1978). *Macromol. Rev.* **13,** 219.

Berkovitch–Yellin, Z., and Leiserowitz, L. (1977). *Acta Crystallogr., Sect. B* **B33,** 3657.

Bertault, T. M., Fave, J. L., and Schott, M. (1979). *Chem. Phys. Lett.* **62,** 161.

Bhattacharjee, H. R., Preziosi, A. F., and Patel, G. N. (1980). *J. Chem. Phys.* **73,** 1478.

Bloor, D. (1980). *Springer Lect. Notes Phys.* **113,** 14.

Bloor, D. (1982). *Dev. Cryst. Polym.* **1,** 151.

Bloor, D. Ando, D. J., Preston, F. H., and Stevens, G. C. (1974). *Chem. Phys. Lett.* **24,** 407.

Bloor, D., and Chance, R. R., eds. (1985). "Polydiacetylenes: Synthesis, Structure and Electronic Properties." Martinus Nijhoff Dordrecht, Netherlands.

Bloor, D., Fisher, D. A., Batchelder, D. N., Kennedy, R., Cottle, A. C., Lewis, W. F., and Hurtshouse, M. B. (1979). *Mol. Cryst. Liq. Cryst.* **52,** 83.

Bloor, D., and Hubble, C. L. (1978). *Chem. Phys. Lett.* **56,** 89.

Bloor, D., Koski, L., Stevens, G. C., Preston, F. H., and Ando, D. J. (1975). *J. Mater. Sci.* **10,** 1678.

Bloor, D., and Preston, H. F. (1976). *Phys. Status Solidi A* **37,** 427.

Bloor, D., and Preston, H. F. (1977). *Phys. Status Solidi A* **39,** 607.

Brédas, J. L., Chance, R. R., Silbey, R., Nicolas, G., and Durand, (1981). *J. Chem. Phys.* **75,** 255.

Brillante, A., Pockrand, I., Philpott, M. R., and Swalen, J. D. (1978). *Chem. Phys. Lett.* **57,** 395.

Bubeck. C., Sixl, H., Bloor, D., and Wegner, G. (1979). *Chem. Phys. Lett.* **63,** 574.

Bubeck, C., Tieke, B., and Wegner, G. (1982). *Ber. Bunsenges. Phys. Chem.* **86,** 495, 499.

Busing, W. R., and Levy, H. A. (1964). *Acta Crystallogr.* **17,** 142.

Cade, N. A., and Movaghar, B. (1982). *J. Phys. C* **16,** 539.

Cadiot, P., and Chodkiewidz, W. (1969). *In* "Chemistry of Acetylenes" (H. G. Viehe, ed.), p. 597. Dekker, New York.

Chance, R. R. (1980). *Macromolecules* **13,** 396.

Chance, R. R., and Baughman, R. H. (1976). *J. Chem. Phys.* **64,** 3889.

Chance, R. R., Baughman, R. H., Müller, H., and Eckhardt, C. J. (1977). *J. Chem. Phys.* **67,** 3616.

Chance, R. R., Patel, G. N., and Witt, J. D. (1979). *J. Chem. Phys.* **71,** 206.

Chance, R. R., Shand, M. L., Hogg, C., and Silbey, R. (1980a). *Phys. Rev. B* **22,** 3540.

Chance, R. R., Yee, K. C., Baughman, R. H., Eckhardt, H., and Eckhardt, C. J. (1980b). *J. Polym. Sci., Polym. Phys. Ed.* **16,** 1651.

Chance, R. R., Washabaugh, M. W., and Hupe, D. J. (1985). *In* "Polydiacetylenes" (D. Bloor and R. R. Chance, eds.), pp. 239–256. Martinus Nijhoff, Dordrecht, Netherlands.

Cojan, C., Agrawal, G. P., and Flytzanis, C. (1977). *Phys. Rev.* **B15,** 909.

Cottle, A. C. (1980). Cited in Williams *et al.* (1980).

Cottle, A. C., Lewis, W. F., and Batchelder, D. N. (1978). *J. Phys. C* **11,** 605.

Cruickshank, D. W. J. (1956). *Acta Crystallogr.* **9,** 757.

Davydov, A. S. (1962). "Theory of Molecular Excitons," McGraw–Hill, New York.

Davydov, A. S. (1971). "Theory of Molecular Excitons," Plenum Press, New York.

Day, D., and Lando, J. B. (1978). *J. Polym. Sci., Polym. Phys. Ed.* **16,** 1009.

Dinur, U., and Karplus, M. (1982). *Chem. Phys. Lett.* **88,** 171.

Dudley, M., Sherwood, J. N., Ando, D. J., and Bloor, D. (1985). *In* "Polydiacetylenes" (D. Bloor and R. R. Chance, eds.), pp. 87–92. Martinus Nijhoff, Dordrecht, Netherlands.

Eckhardt, H., Eckhardt, C. J., and Yee, K. C. (1979). *J. Chem. Phys.* **70,** 5498.

Egri, I. (1985). *Phys. Rep.* **119,** 363.

Eichele, H., Herath, E., and Kröhnke, C. (1980). *Chem. Phys. Lett.* **71,** 211.

Enkelmann, V. (1977). *Acta Crystallogr., Sect. B* **B33,** 2842.

Enkelmann, V. (1984). *Adv. Polym. Sci.* **63,** 91.

Enkelmann, V., and Lando, J. B. (1978). *Acta Crystallogr.* **B34,** 2352.

Enkelmann, V., and Schleier, G. (1980). *Acta Crystallogr., Sect. B* **B36,** 1954.

Enkelmann, V., Schleier, G., Wegner, G., Eichele, H., and Schwoerer, M. (1977). *Chem. Phys. Lett.* **52,** 314.

Enkelmann, V., Leyrer, R. J., and Wegner, G. (1979). *Makromol. Chem.* **180,** 1787.

Enkelmann, V., Leyrer, R. J., Schleier, G., and Wegner, G. (1980). *J. Mater. Sci.* **15,** 168.

Exarhos, G. J., Risen, W. M., Jr., and Baughman, R. H. (1978). *J. Am. Chem. Soc.* **98,** 481.

Fincher, C. R., Jr., Chen, C. E., Heeger, A. J., Macdiarmid, A. G., and Hastings, J. B. (1982). *Phys. Rev. Lett.* **48,** 100.

Finkelman, H., and Rehage, G. (1984). *Adv. Polym. Sci.* **60/61,** 99.

Fukuyama, I., Kuchitsu, K., and Morino, Y. (1969). *Bull. Chem. Soc. Jpn.* **42,** 379.

Garito, A. F., McGhie, A. R., and Kalyanaraman, P. S. (1979). *In* "Molecular Metals" (W. E. Hatfield, ed.), p. 255. Plenum, New York.

Gros, L., Ringsdorf, H., and Schupp, H. (1981). *Angew. Chem.* **93,** 311, and references cited therein

Hädicke, E., Mez, E. C., Krauch, C. H., Wegner, G., and Kaiser, J. (1971). *Angew. Chem.* **83,** 253; *Angew. Chem. Int. Ed. Eng.* **10,** 266.

Hay, A. S. (1962). *J. Org. Chem.* **27,** 3320.

Hood, R. J., Müller, H., Eckhardt, C. J., Chance, R. R., and Yee, K. C. (1978). *Chem. Phys. Lett.* **54,** 483.

Horie, C. (1959). *Prog. Theor. Phys.* **21,** 113.

Hudson, B. S., Kohler, B. E., and Schulten, K. (1982). *In* "Excited States" (E. C. Lim, ed.), Vol. 6. Academic Press, New York.

Iqbal, Z., Chance, R. R., and Baughman, R. H. (1977). *J. Chem. Phys.* **66,** 5520.

Irngartinger, H., and Jäger, H. U. (1976). *Angew. Chem. Int. Ed. Engl.* **15,** 562.

Jankowiak, R., Kalinowski, J., Reimer, B., and Bässler, H. (1978). *Chem. Phys. Lett.* **54,** 483.

Johnston, D. S., Sanghera, S., Pons, M., and Chapman, D. (1980). *Biochim. Biophys. Acta* **602,** 57.

Kaiser, J., Wegner, G., and Fischer, E. W. (1972). *Isr. J. Chem.* **10,** 157.

Kajzar, F., and Messier, J. (1985). *In* "Polydiacetylenes" (D. Bloor and R. R. Chance, eds.), pp. 325–333. Martinus Nijhoff, Dordrecht, Netherlands.

Karpfen, A. (1980). *J. Phys. C* **13,** 5673.

Kawaoka, K. (1976). *Chem. Phys. Lett.* **37,** 561.

Kobelt T. D., and Paulus, E. F. (1974). *Acta Crystallogr., Sect. B* **B30**, 232.

Kröhnke, C., Enkelmann, V., and Wegner, G. (1980). *Chem. Phys. Lett.* **71**, 38.

Lacey, R. J., Batchelder, D. N., and Pitt, G. D. (1984). *J. Phys. C* **17**, 4529.

Lando, J. B. (1985). *In* "Polydiacetylenes" (D. Bloor and R. Chance, eds.), pp. 363–370. Martinus Nijhoff, Dordrecht.

Lauritzen, J. I., and Hoffman, J. D. (1960). *J. Res. Natl. Bur. Stand. Sect. A* **64**, 79.

Lequime, M., and Hermann, J. P. (1977). *Chem. Phys.* **26**, 431.

Leyrer, R. J., and Wegner, G. (1979). *Ber. Bunsenges. Phys. Chem.* **83**, 470.

Leyrer, R. J., Wegner, G., and Wettling, W. (1978). *Ber. Bunsenges. Phys. Chem.* **82**, 697.

Lieser, G., Tieke, B., and Wegner, G. (1980). *Thin Solid Films* **68**, 77.

Lim, K. C., Kapitulnik, A., Zacher, R., Casalnuovo, S., Wudl, F., and Heeger, A. J. (1985). *In* "Polydiacetylenes" (D. Bloor and R. R. Chance, eds.), pp. 257–290. Martinus Nijhoff, Dordrecht, Netherlands.

Lochner, K., Bässler, H., Tieke, B., and Wegner, G. (1978). *Phys. Status Solidi B* **88**, 653.

Lochner, K., Bässler, H., Sowa, H., and Ahsbahs, H. (1980). *Chem. Phys.* **52**, 179.

Lochner, K., Bässler, H., Sebastian, L., Weiser, G., Wegner, G., and Enkelmann, V. (1981). *Chem. Phys. Lett.* **78**, 366.

Longuet–Higgins, H. C., and Pople, J. A. (1957). *J. Chem. Phys.* **27**, 192.

McGhie, A. R., Lipscomb, G. F., Garito, A. F., Desai, K. N., Kalyanaraman, P. S. (1981). *Makromol. Chem.* **182**, 965.

Müller, H., and Eckhardt, C. J. (1978). *J. Chem. Phys.* **67**, 5386.

Müller, H., Eckhardt, C. J., Chance, R. R., and Baughman, R. H. (1977). *Chem. Phys. Lett.* **50**, 22.

Müller, M. A., and Wegner, G. (1984). *Makromol. Chem.* **185**, 1727.

Murashov, A. A., Silinsh, E. A., and Bässler, H. (1982). *Chem. Phys. Lett.* **93**, 148.

Niederwald, H., and Schwoerer, M. (1983). *Z. Naturforsch. A* **38a**, 749.

Orrit, M., Aslangul, C., and Kottis, P. (1982). *Phys. Rev. B* **25**, 7263.

Patel, G. N. (1978a). *Polym. Prepr., Am. Chem. Soc., Div. Polym. Chem.* **19** (2), 154.

Patel, G. N. (1978b). *J. Polym. Sci., Polym. Lett. Ed.* **16**, 607.

Patel, G. N. (1981). *Macromolecules* **14**, 1170.

Patel, G. N., Chance, R. R., and Witt, J. D. (1979). *J. Chem. Phys.* **70**, 4387.

Philpott, M. R. (1973). *Adv. Chem. Phys.* **23**, 227–341.

Philpott, M. R. (1977). *Chem. Phys. Lett.* **50**, 18.

Philpott, M. R. (1980). *Annu. Rev. Phys. Chem.* **31**, 97.

Plachetta, C., and Schulz, R. C. (1982). *Makromol. Chem., Rapid Commun.* **3**, 815.

Plachetta, C., Rau, N. O., Hauck, A., and Schulz, R. C. (1982). *Makromol. Chem., Rapid Commun.* **3**, 249.

Pope, M., and Swenberg, C. E. (1982). "Electronic Processes in Organic Crystals." Oxford Univ. Press (Clarendon), London and New York.

Pugh, D. (1973). *Mol. Phys.* **26**, 1297.

Rawiso, M. *et al.* (1986). To be published.

Rehwald, W., Vonlanthen, A., and Meyer, W. (1983). *Phys. Status Solidi A* **75**, 219.

Reimer, B., and Bässler, H. (1975). *Phys. Status Solidi A* **32**, 435.

Reimer, B., and Bässler, H. (1978). *Chem. Phys. Lett.* **55**, 315.

Rei Vilar, M. (1985). Thesis, University of Paris VII (unpublished).

Rice, S. A., and Jortner, J. (1965). *In* "Physics of Solids at High Pressures" (C. T. Tomizuka and R. M. Emrick, eds.), pp. 63–168.

Ritsko, J. J., Crecelius, G., and Fink, J. (1982). *Phys. Rev. B* **27**, 4902.

Sandman, D. J., Carter, G. M., Chen, Y. J., Elman, B. S., Thakur, M. K., and Tripathy, S. K. (1985). *In* "Polydiacetylenes" (D. Bloor and R. R. Chance, eds.), pp. 299–316. Martinus Nijhoff, Dordrecht, Netherlands.

Sauteret, C., Herrmann, J. P., Frey, R., Pradère, F., Ducuing, J., Baughman, R. H., and Chance, R. R. (1976). *Phys. Rev. Lett.* **36**, 956.

Schermann, W., Wegner, G., Williams, J. O., and Thomas, J. M. (1975). *J. Polym. Sci., Polym. Phys. Ed.* **13**, 753.

Schott, M., Batallan, F., and Bertault, M. (1978). *Chem. Phys. Lett.* **53**, 443.

Schulten, K., Ohmine, I., and Karplus, M. (1976). *J. Chem. Phys.* **64**, 4422.

Schwoerer, M., and Niederwald, H. (1984). *In* "Photoreaktive Festkörper" (H. Sixl, J. Friedrich, and L. Bräuchle, eds.), p. 291. M. Wahl Verlag, Karlsruhe.

Sebastian, L. (1980). Thesis, University of Marburg (unpublished).

Sebastian, L., and Weiser, G. (1979). *Chem. Phys. Lett.* **64**, 396.

Sebastian, L., and Weiser, G. (1981a). *Phys. Rev. Lett.* **46**, 1156.

Sebastian, L., and Weiser, G. (1981b). *Chem. Phys.* **62**, 447.

Shand, M., Chance, R. R., Lepostollec, M., and Schott, M. (1982). *Phys. Rev. B* **25**, 4431.

Siddiqui, A. S. (1980). *J. Phys. C* **13**, 2147.

Siegel, D., Sixl, H., Enkelmann, V., and Wenz, G. (1982). *Chem. Phys.* **72**, 201.

Sixl, H. (1984). *Adv. Polym. Sci.* **63**, 49.

Sixl, H. (1985). *In* "Polydiacetylenes" (D. Bloor and R. R. Chance, eds.), pp. 41–65. Nijhoff, Dordrecht, Netherlands.

Spinat, P., Brouty, C., Whuler, A., and Sichere, M. C. (1985). *Acta Crystallogr., Part C* **C41**, 1452.

Suhai, S. (1984). *Phys. Rev. B* **29**, 4570.

Tieke, B., Wegner, G., Naegele, D., and Ringsdorf, H. (1976). *Angew. Chem., Int. Ed. Engl.* **12**, 764.

Tieke, B., Graf, H. J., Wegner, G., Naegele, B., Ringsdorf, H., Banerjie, A., Day, D., and Lando, J. B. (1977). *Colloid Polym. Sci.* **255**, 521.

Tieke, B., Lieser, G., and Wegner, G. (1979). *J. Polym. Sci., Polym. Chem. Ed.* **17**, 1631.

Tieke, B., Enkelmann, V., Kapp, H., Lieser, G., and Wegner, G. (1981). *J. Macromol. Sci., Chem. A* **15**, 1045.

Tokura, Y., Mitani, T., and Koda, T. (1980). *Chem. Phys. Lett.* **75**, 324.

Tokura, Y., Owaki, Y., Koda, T., and Baughman, R. H. (1984). *Chem. Phys.* **88**, 437.

Wegner, G. (1969). *Z. Naturforsch.* **24B**, 824.

Wegner, G. (1971a). *J. Polym. Sci., Polym. Lett. Ed.* **9**, 133.

Wegner, G. (1971b). *Makromol. Chem.* **145**, 85.

Wegner, G. (1972). *Makromol. Chem.* **154**, 35.

Wegner, G. (1984). *Makromol. Chem., Suppl.* **6**, 347.

Wegner, G., Munoz–Escalona, A., and Fischer, E. W. (1975). *Makromol. Chem., Suppl.* **1**, 521.

Weiser, G., and Sebastian, L. (1985). *In* "Polydiacetylenes" (D. Bloor and R. R. Chance, eds.), pp. 213–222. Martinus Nijhoff, Dordrecht, Netherlands.

Wenz, G., and Wegner, G. (1982). *Makromol. Chem., Rapid Commun.* **3**, 231.

Wenz, G., Müller, M. A., Schmidt, M., and Wegner, G. (1984). *Macromolecules* **17**, 837.

Whangbo, M. H., Hoffmann, R., and Woodward, R. B. (1979). *Proc. R. Soc. London, Ser. A* **366**, 23.

Williams, R. L., Ando, D. J., Bloor, D., and Hurtshouse, M. B. (1980). *Polymer* **21**, 1269.

Wilson, E. G. (1975). *J. Phys. C* **8**, 727.

Wilson, E. G. (1985). *In* "Polydiacetylenes (D. Bloor and R. R. Chance, eds.), pp. 155–164. Martinus Nijhoff, Dordrecht, Netherlands.

Wunderlich, B. (1968). *Adv. Polym. Sci.* **5**, 568.

Yarkony, D. R. (1978). *Chem. Phys.* **33**, 171.

Yee, K. C., and Chance, R. R. (1978). *J. Polym. Sci., Polym. Phys. Ed.* **16**, 431.

Young, R. J. (1985). *In* "Polydiacetylenes" (D. Bloor and R. R. Chance, eds.), pp. 335–362. Martinus Nijhoff, Dordrecht, Netherlands.

Zachmann, H. G. (1969). *Kolloid-Z. Z. Polym.* **231**, 504.

Chapter III-2

Cubic Effects in Polydiacetylene Solutions and Films

F. KAJZAR and J. MESSIER

Commissariat à l'Energie Atomique,
Institut de Recherche Technologique et de Developpement Industriel,
Département d'Electronique et d'Instrumentation Nucléaire,
Laboratoire d'Etudes et Recherches Avancées,
Centre d'Etudes Nucléaires de Saclay,
91191 Gif-sur-Yvette Cedex, France

I. INTRODUCTION

A. Polarization in One-Dimensional Media

The macroscopic polarization P of a dielectric can be expanded in the external electric field power series

$$P = \chi^{(1)} : E + \chi^{(2)} : EE + \chi^{(3)} : EEE + \cdots \tag{1}$$

where the expansion coefficients $\chi^{(n)}$ are $(n + 1)$-rank three-dimensional tensors. For species with centrosymmetric structure, like polydiacetylenes, all odd-rank susceptibilities are zero $[\chi^{(2n)} \equiv 0]$. The case of noncentrosymmetric molecules is treated elsewhere (Zyss and Chemla, Vol. 1, Chapter II-1). Similar expansion to Eq. (1) is obvious for the microscopic molecular polarization p,

$$p = \alpha : E + \beta : EE + \gamma : EEE + \cdots \tag{2}$$

51

where α, β, and γ are molecular susceptibilities. In isotropic media α and γ are related to the macroscopic ones through local field factors

$$\chi^{(1)} = \alpha L_1 N$$
$$\chi^{(3)} = \gamma L_3 N \tag{3}$$

where L_i are the local field factors, which in Lorentz approximation are given by

$$L_i = \left[\frac{n^2 + 2}{3}\right]^{i+1} \tag{4}$$

and N is the density of molecules.

The electronic polarization of a molecule is connected with the delocalization of electronic cloud: the most important contribution coming from the outer electron shell. The calculations done by Sewell (1949) for the hydrogen atom give

$$\alpha = \tfrac{9}{2} a_0^3 \tag{5}$$
$$\gamma = \tfrac{3555}{16} a_0^7 \tag{6}$$

where a_0 is the Bohr radius.

Equations (5) and (6) show the importance of electron delocalization on hyperpolarizability values. In fact, the polarizability is largest when the delocalization is in one dimension. The calculations in the Unsold approximation give (Ducuing, 1977)

$$\alpha \propto L^4 / \mathcal{N} \tag{7}$$
$$\gamma \propto L^{10} / \mathcal{N}^3 \tag{8}$$

where L is the delocalization length and $\mathcal{N}$ the number of electrons.

For a one-dimensional system, $\mathcal{N} \propto L$; thus $\alpha \propto L^3$ and $\gamma \propto L^7$. For a two-dimensional system $\mathcal{N} \propto L^2$ and correspondingly $\alpha \propto L^2$ and $\gamma \propto L^4$. This shows the importance of delocalization and of dimensionality in polarizability of molecules.

As discussed by Ducuing (1977), the dependence on L in the Unsold approximation is probably overestimated. In fact, the calculations for a one-dimensional free-electron gas (Rustagi and Ducuing, 1974) give a slightly weaker dependence of α and γ on L ($\alpha \propto L^3$ and $\gamma \propto L^5$). Similarly, detailed calculations of Agrawal et al. (1978) done for polydiacetylenes in a tight-binding approximation give also a weaker dependence on L ($\gamma \propto L^6$).

B. Origin of Cubic Nonlinearities in Polydiacetylenes

Polydiacetylenes with the general formula $R{-}C{-}C{\equiv}C{-}C{-}R'$ are a model system of conjugated π electrons in one dimension. Their principal advantage is topochemical polymerization under ultraviolet (UV), γ irradiation, and/or heating in a variety of different structural forms: amorphous, semicrystalline, and crystalline thin films, Langmuir–Blodgett mono- and multilayers, and single crystals (for a review of structural properties of polydiacetylenes, see Schott and Wegner, Chapter III-1). Some of these polymers, with appropriate side groups, are soluble in a large number of organic solvents.

First calculations of cubic hyperpolarizabilities of polydiacetylenes were done by Ducuing (1977) in the framework of Penn's one-dimensional model, which gave

$$\chi^{(3)} \propto \frac{D^6}{(E_F d)^3} \tag{9}$$

where D is a delocalization parameter, d the interatomic spacing, and E_F the Fermi energy. Similar results were obtained by Agrawal $et\ al.$ (1978) in a tight-binding approximation using the Genkin and Mednis (1968) approach for nonlinear susceptibilities. Agrawal $et\ al.$ (1978) found $\chi^{(3)}$ on the order of 10^{-11} esu for polymers, depending slightly on side groups, and two orders of magnitude smaller values for monomers.

In their calculations, Agrawal $et\ al.$ (1978) assume $a\ priori$ that the first optical transition is a band-to-band transition, whereas low-temperature reflectance study (Sebastian and Weiser, 1981), photoconduction (Lochner $et\ al.$, 1978), as well as electric-field-induced second-harmonic generation (EFISHG) measurements (Chollet $et\ al.$, 1985 and 1986) on polydiacetylene (PDA) thin films indicate that the first optically allowed state is an excitonic band lying in blue form (see Fig. 1) of the polymer at about 15,600 cm^{-1} above the valence band, whereas the conduction band is situated at about 18,900 cm^{-1}. Within such a localized picture, one can assume that the quantum-mechanical formulas derived for isolated atoms are valid and the largest tensor

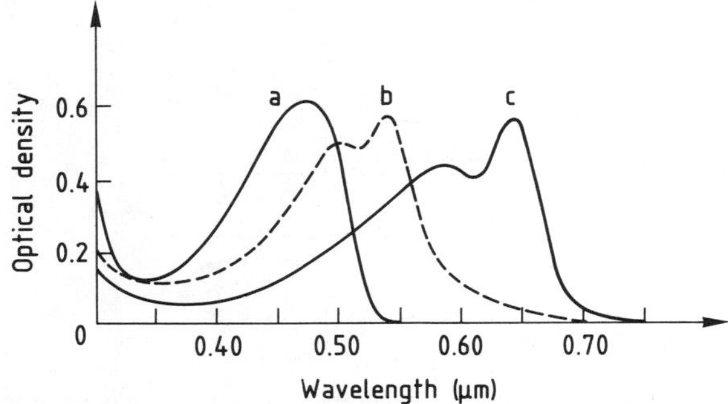

Fig. 1. Optical absorption spectra of (a) yellow solution of polymer, (b) red, and (c) blue forms of LB films.

component responding for THG is given by (Ward, 1965; Langhof *et al.*, 1972; Hanna *et al.*, 1979)

$$\chi_{xxxx}^{(3)}(-3\omega;\omega,\omega,\omega)\alpha\frac{N}{h^3}\sum_{gnmn'}\rho(g)\Omega_{gn}\Omega_{nm}\Omega_{mn'}\Omega_{n'g}$$

$$\times\left[\frac{1}{(E_{ng}-3\omega)(E_{mg}-2\omega)(E_{n'g}-\omega)}+\frac{1}{(E_{ng}+\omega)(E_{mg}-2\omega)(E_{n'g}-\omega)}\right.$$

$$\left.+\frac{1}{(E_{ng}+\omega)(E_{mg}+2\omega)(E_{n'g}-\omega)}+\frac{1}{(E_{ng}+\omega)(E_{mg}+2\omega)(E_{n'g}+3\omega)}\right] \quad (10)$$

where $\rho(g)$ is the density matrix element of fundamental state, E_{ij} the energy difference between states j and i in h units, and Ω_{ij} the transition matrix elements.

Equation (10) shows a frequency dependence of cubic susceptibility, with the possibility of resonance enhancement depending on the electronic structure of polydiacetylenes. Because the dipolar-moment operators have odd parity, the state m must have the same symmetry as the fundamental one, and the states n and n' an opposite one (see Fig. 2). This means that one can observe a two-photon resonance in the polymer transparency region, as is the case. Moreover, due to the high value of terms such as Ω_{nm} in a one-dimensional molecule, high values of $\chi^{(3)}$ are expected for polydiacetylenes. In fact, first, third-harmonic generation (THG) measurements performed by Sauteret *et al.* (1976) on a PTS single crystal confirmed these expectations. The measured value $\chi_{xxxx}^{(3)}(-3\omega;\omega,\omega,\omega) = (8.5 \pm 5) \times 10^{-10}$ esu is the highest observed for organic materials and is comparable to that of inorganic semiconductors with the principal advantage of electronic origin and consequently fast response time.

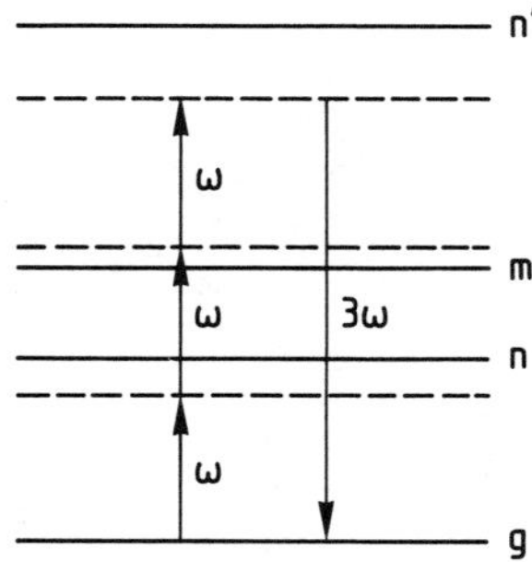

Fig. 2. Schematic representation of third-harmonic generation process. Solid lines represent excited states, and broken lines represent virtual states.

In this chapter we discuss third (THG) and electric-field-induced second harmonic generation, (EFISHG), as well as four-wave mixing experiments (FWM) on polydiacetylenes. Other cubic effects like nonlinear index variation in these materials are discussed by Carter *et al.* (Chapter III-3).

II. THEORY

A. Nonlinear Propagation and Harmonic Generation in Stratified Media

In this section we give some fundamental formulas for harmonic generation (THG and EFISHG) leading to determination of corresponding cubic susceptibilities.

A monochromatic plane wave

$$E^{\omega}(\mathbf{r}t) = \tfrac{1}{2}[E^{\omega}(\mathbf{r})\exp(-i\omega t) + cc] \tag{11}$$

generates in a dielectric a nonlinear polarization at harmonic frequency ω_{H} given by

$$P_{\mathrm{NL}}(\mathbf{r}t) = \tfrac{1}{2}[P_{\mathrm{NL}}(\mathbf{r})\exp(-i\omega_{\mathrm{H}}t) + cc] \tag{12}$$

where for cubic processes considered in this chapter,

$$P_{\mathrm{NL}}(\mathbf{r}) = B\chi^{(3)}(-3\omega; \omega, \omega, \omega)(E^{\omega})^3 \qquad B = \tfrac{1}{4} \quad \text{for THG} \tag{13}$$

$$P_{\mathrm{NL}}(\mathbf{r}) = B\chi^{(3)}(-2\omega; \omega, \omega, 0)(E^{\omega})^2 E_0 \qquad B = \tfrac{3}{2} \quad \text{for EFISHG} \tag{14}$$

The factor B in Eqs. (13) and (14) arises from permutation symmetry properties of $\chi^{(3)}$ and from the electric-field definition [Eq. (11)]. The dc field E_0 in Eq. (14) is its maximum value in the present convention.

The nonlinear polarization P_{NL} is a source term in the Maxwell equation at harmonic frequency ω_{H}:

$$\nabla^2 E^{\omega_{\mathrm{H}}} - \frac{\varepsilon(\omega_{\mathrm{H}})}{c^2}\left(\frac{\partial^2 E^{\omega_{\mathrm{H}}}}{\partial t^2}\right) = \frac{4\pi}{c^2}\left(\frac{\partial P_{\mathrm{NL}}}{\partial t^2}\right) \tag{15}$$

The solution $E^{\omega_{\mathrm{H}}}$ of Eq. (15) is a sum of two solutions with (bound wave) and without (free wave) source: both are connected by boundary conditions ensuring electric field continuity through different interfaces.

In the case when the transfer of power from fundamental to harmonic wave is sufficiently small, the propagation of fundamental wave is not disturbed by generated harmonic waves and the bound wave can be determined independently from free waves. From the fact that Eq. (15) is linear in P_{NL}, it follows that the bound waves can be determined independently in all of n nonlinear media ($n \geqslant 1$). Thus it is possible to determine for every bound wave the associated free waves. In this case the resulting harmonic field will be a linear superposition of these free waves. The problem of nonlinear propagation in n stratified media can be thus reduced to one medium immersed between two linear media, and the resulting harmonic field will be equal to the sum of harmonic fields generated in each medium separately.

1. Third Harmonic Generation in a Plane-Parallel Slab

For an incident fundamental wave with frequency ω given by Eq. (11) and neglecting its multiple reflections, the bound wave at $\omega_H = 3\omega$ in medium 2 (see Fig. 3) is given by

$$E_{2b}^{3\omega}(\mathbf{r}, t) = \tfrac{1}{2}[E_{2b} \exp i(\mathbf{k}_{2b} \cdot \mathbf{r} - \omega t) + cc] \tag{16}$$

where

$$E_{2b} = \frac{\pi \chi^{(3)}(E_{2t}^{\omega})^3}{(k_{2b})^2 - (k_2^{3\omega})^2}\left(\frac{\omega}{c}\right)^2 = \frac{\pi \chi^{(3)}(E_{2t}^{\omega})^3}{\Delta\varepsilon} \tag{17}$$

where $E_{2t}^{\omega} = E^{\omega} t_{12}^{\omega}$ is a fundamental wave in medium 2;

$$t_{12}^{\omega} = 2n_2^{\omega}/(n_1^{\omega} + n_2^{\omega})$$

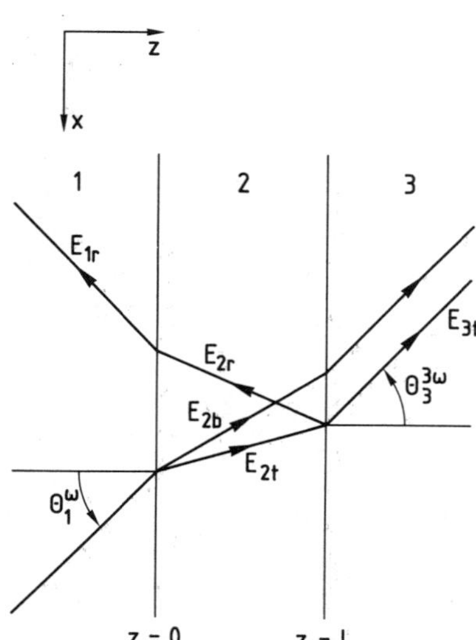

Fig. 3. Harmonic generation in a nonlinear medium (2) immersed between two linear media (1 and 3). Arrows show harmonic wave vectors. Electric fields are perpendicular to the figure plane

is the transmission factor between media 1 and 2 and $\Delta\varepsilon = \varepsilon_2^\omega - \varepsilon_2^{3\omega}$ is the dielectric constant dispersion, the subscript 2 refering to the considered medium. The associated free waves are as follows:

$$\tfrac{1}{2}[E_{1r}^{3\omega}\exp i(\mathbf{k}_{1r}^{3\omega}\cdot\mathbf{r} - 3\omega t) + cc] \qquad \text{free wave transmitted from 2 to 1}$$

$$\tfrac{1}{2}[E_{2t}^{3\omega}\exp i(\mathbf{k}_{2t}^{3\omega}\cdot\mathbf{r} - 3\omega t) + cc] \qquad \text{free wave propagating in medium 2 in } Z \text{ direction}$$

$$\tfrac{1}{2}[E_{2t}^{3\omega}\exp i(\mathbf{k}_{2t}^{3\omega}\cdot\mathbf{r} - 3\omega t) + cc] \qquad \text{free wave propagating in medium 2 in } -Z \text{ direction} \tag{18}$$

$$\tfrac{1}{2}\{E_{3t}^{3\omega}\exp i[\mathbf{k}_{3t}^{3\omega}\cdot(\mathbf{r} - \mathbf{l}) - 3\omega t] + cc\} \qquad \text{free wave transmitted to medium 3}$$

where vector $\mathbf{l}$ points in the z direction and $|\mathbf{l}| = l$ is the medium thickness.

The continuity of tangential components of wave vectors at interfaces implies that the E_{3t} wave propagates in the same direction as the fundamental one.

From the continuity of the electric field at interfaces 1–2 and 2–3, one obtains the following equations for electric-field amplitudes:

$$\begin{aligned} E_{1r} &= E_{2t} + E_{2r} + E_{2b} \\ -N_1^{3\omega}E_{1r} &= N_2^{3\omega}(E_{2t} - E_{2r}) + N_2^\omega E_{2b} \end{aligned} \qquad z = 0 \tag{19}$$

$$\begin{aligned} E_{2t}e^{i\psi_3} + E_{2r}e^{-i\psi_3} + E_{2b}e^{i\psi_1} &= E_{3t} \\ N_2^{3\omega}(E_{2t}e^{i\Psi_3} - E_{2r}e^{-i\psi_3}) + N_2^\omega E_{2b}e^{i\psi_1} &= N_3^{3\omega}E_{3t} \end{aligned} \qquad z = l \tag{20}$$

where

$$\psi_m = 3\omega N_2^{m\omega}l/c \qquad m = 1, 3$$

and

$$N_j^{m\omega} = \frac{c}{3\omega}(\mathbf{k}_j^{m\omega}\cdot\mathbf{r}) = n_j^{m\omega}\cos\theta_j^{m\omega} \tag{21}$$

By solving simultaneously this set of equations, one obtains

$$E_{3t} = A_0\exp i(\mathbf{k}_{2t}^{3\omega}\cdot\mathbf{l})\{A_1[\exp i(\mathbf{k}_{2b} - \mathbf{k}_{2t}^{3\omega})\cdot\mathbf{l} - 1]$$
$$+ A_2[\exp i(\mathbf{k}_{2b} + \mathbf{k}_{2t}^{3\omega})]\cdot\mathbf{l} - 1\}/D \tag{22}$$

$$E_{1r} = A_0\exp i[(\mathbf{k}_{2b} + \mathbf{k}_{2t}^{3\omega})\cdot\mathbf{l}](A_3\{\exp[-i(\mathbf{k}_{2b} - \mathbf{k}_{2t}^{3\omega})\cdot\mathbf{l}] - 1\}$$
$$+ A_4\{\exp[-i(\mathbf{k}_{2b} + \mathbf{k}_{2t}^{3\omega})\cdot\mathbf{l}] - 1\})/D \tag{23}$$

where $A_0 = E_{2b}$ [cf. Eq. (17)] and

$$A_1 = \frac{N_2^{3\omega} + N_2^{\omega}}{N_3^{3\omega} + N_2^{3\omega}}, \qquad A_2 = \left(\frac{N_2^{3\omega} - N_2^{\omega}}{N_3^{3\omega} + N_2^{3\omega}}\right)\left(\frac{N_1^{3\omega} - N_2^{3\omega}}{N_1^{3\omega} + N_2^{3\omega}}\right) \tag{24}$$

$$A_3 = \left(\frac{N_3^{3\omega} - N_2^{3\omega}}{N_1^{3\omega} + N_2^{3\omega}}\right)\left(\frac{N_2^{\omega} + N_2^{3\omega}}{N_3^{3\omega} + N_2^{3\omega}}\right), \qquad A_4 = \frac{N_2^{3\omega} - N_2^{\omega}}{N_1^{3\omega} + N_2^{3\omega}} \tag{25}$$

$$D = 1 - \left(\frac{N_1^{3\omega} - N_2^{3\omega}}{N_1^{3\omega} + N_2^{3\omega}}\right)\left(\frac{N_3^{3\omega} - N_2^{3\omega}}{N_3^{3\omega} + N_2^{3\omega}}\right)\exp[i(6\omega l N_2^{3\omega}/c)] \tag{26}$$

Equations (22) and (23) show the existence of two coherence lengths: for the transmitted wave (E_{3t}), $l_c = |\lambda/6(n_2^{\omega} - n_2^{3\omega})|$, and for the reflected wave (E_{1r}), $l_c^r = |\lambda/6(n_2^{\omega} + n_2^{3\omega})|$. Generally, $l_c \gg l_c^r$ and if $l > l_c$, then the amplitude of the reflected field E_{1r} is much smaller than that of the transmitted wave E_{3t}. However, for very thin nonlinear media, like thin films, $l < l_c$ and E_{1r} can be of the same size of magnitude as E_{3t}.

Generally, A_2 [Eq. (24)] is very small compared with A_1, and the second term in Eq. (22) can be neglected. Similarly, D [Eq. (26)] is close to 1.

2. Multiple Reflections of a Fundamental Wave

Equations (22)–(26) take into account multiple reflections of harmonic wave but do not include those of the fundamental wave. These can be important, especially on interfaces of nonlinear medium–vacuum (or air), where the reflection coefficient $r = (n_2^{\omega} - 1)/(n_2^{\omega} + 1)$ is not negligible.

The fundamental wave in medium 2 can be described as a superposition of two waves, E_{2i}^{ω} and E_{2r}^{ω}, given by

$$E_{2i}^{\omega} = \frac{1}{2}\left\{E_{2t}^{\omega}\frac{1}{1 - r^2 e^{i2\psi}}\exp[i(\mathbf{k}_{2i}^{\omega}\cdot\mathbf{r} - \omega t)] + cc\right\} \tag{27}$$

and

$$E_{2r}^{\omega} = \frac{1}{2}\left\{E_{2t}^{\omega}\frac{r e^{i2\psi}}{1 - r^2 e^{i2\psi}}\exp[i(\mathbf{k}_{2r}^{\omega}\cdot\mathbf{r} - \omega t)] + cc\right\} \tag{28}$$

where

$$\mathbf{k}_{2r}^{\omega}\cdot\mathbf{x} = \mathbf{k}_{2i}^{\omega}\cdot\mathbf{x} \qquad \mathbf{k}_{2r}^{\omega}\cdot\mathbf{z} = -\mathbf{k}_{2i}^{\omega}\cdot\mathbf{z} \tag{29}$$

and

$$\psi = \omega N_2^{\omega} l/c \tag{30}$$

Then E_{2t}^{ω} is the amplitude of the fundamental wave transmitted into medium 2.

There exist four bound waves corresponding to different superposition of the fundamental fields

$$E_\alpha^\omega E_\beta^\omega E_\gamma^\omega \qquad (\alpha, \beta, \gamma = 2i \quad \text{or} \quad 2r)$$

whose amplitude is given by

$$E_{2b}^m = \pi \frac{r^m e^{i2m\psi}}{(1 - r^2 e^{i2\psi})^3} D_m \chi^{(3)}(-3\omega; \omega, \omega, \omega)(E_{2t}^\omega)^3 \qquad (t = i \quad \text{or} \quad r) \quad (31)$$

where

$$D_m = 1/[(k_{2b}^m)^2 - (k_2^{3\omega})^2] \tag{32}$$

$$\mathbf{k}_{2b}^m = (3 - m)\mathbf{k}_{2i}^\omega + m\mathbf{k}_{2r}^\omega \tag{33}$$

and m is the number of E_{2r}^ω fields ($m = 0, 1, 2, 3$) intervening in Eq. (31).

From Eqs. (33) and (29) one can see that $|D_1| = |D_2| \ll |D_0| = |D_3|$.

The continuity of tangential components of different reflected and transmitted beam wave vectors

$$\mathbf{k}_{2b} \cdot \mathbf{x} = (\mathbf{k}_{1r}^{3\omega})_m \cdot \mathbf{x} = (\mathbf{k}_{3t}^{3\omega})_m \cdot \mathbf{x} \tag{34}$$

imposes the same propagation direction for all harmonic waves in medium 1 and in medium 3 (see Fig. 3). In medium 3 the harmonic waves propagate in the direction of the fundamental wave, whereas in medium 1 the propagation direction of the E_{1r} wave makes an angle $2\theta_1^\omega$ with the incident wave (θ_1^ω incidence angle) (if media 1 and 3 are vacuum).

The amplitude of the mth bound wave ($m > 1$) [Eq. (31)] is an oscillating function of phase ψ that depends on medium thickness and incidence angle [cf. Eq. (30)]. Thus the correction to harmonic intensity for multiple reflections ($m \geqslant 1$) will be also an oscillating function with period

$$\Delta\theta = \lambda[(n_2^\omega)^2 - \sin^2 \theta_1^\omega]^{3/2}/n_2^\omega l \sin 2\theta_1^\omega \tag{35}$$

For a thick nonlinear medium, the oscillations are very rapid (e.g., for a silica plate with $l = 1$ mm at $\lambda = 1.064$ μm and $\theta_1^\omega = 5°$, the period of oscillations is $\Delta\theta = 0.7°$), and they will not be resolved by a detection system (finite aperture and beam convergence). In this case the multiple reflections will introduce an average modification of harmonic intensity given by a correction factor

$$R(\theta) = 1 + 9r^4 + O(r^6) \tag{36}$$

This is no longer true for thin films where the period of oscillations is large (ψ varies slowly with incidence angle) and the corrections to harmonic intensity are proportional to $(D_m/D_0)^2 r^{2m}$ for $m = 1, 2, 3$.

For incoherent reflections (small variations in medium thickness will introduce large variations in phase ψ), the corresponding corrections will be of the order of r^6 or r^{12}. The same will happen also for large incidence angle, where (except for very thin films) the reflected and incident waves will not overlap.

Thus, except normal incidence and thin films, the multiple reflections of fundamental beam do not introduce a significant contribution to harmonic intensities and can be neglected.

3. Multiple Reflection of Harmonic Waves

As we mentioned before, Eq. (22) takes into account explicitly multiple reflections of the fundamental wave on interfaces 1–2 and 2–3. Using the argument of surface roughness (medium thickness variation), these effects are systematically neglected (see, e.g., Jerphagnon and Kurz, 1970; Oudar, 1977; Meredith *et al.*, 1983a). It corresponds to setting in Eq. (20) $E_{2r} = 0$ (neglect of all reflected waves from other interfaces than that under consideration). The formulas (22) and (23) for harmonic fields in this case become

$$E_{3t}^{3\omega} = A_0 \exp(i\mathbf{k}_2^{3\omega} \cdot \mathbf{l})\{\alpha_1 \exp[i(\mathbf{k}_{2b} - \mathbf{k}_{2t}^{3\omega})] \cdot \mathbf{l} - \alpha_2\} \tag{37}$$

$$E_{1r} = \alpha_3 A_0 \tag{38}$$

where

$$\alpha_1 = \frac{N_2^{3\omega} + N_2^{\omega}}{N_3^{3\omega} + N_2^{3\omega}}, \qquad \alpha_3 = \frac{N_1^{3\omega} - N_2^{3\omega}}{N_1^{3\omega} + N_2^{3\omega}}$$

$$\alpha_2 = \frac{2N_2^{3\omega}}{N_3^{3\omega} + N_2^{3\omega}}\left(\frac{N_1^{3\omega} + N_2^{3\omega}}{N_1^{3\omega} + N_2^{3\omega}}\right) \tag{39}$$

These formulas do not hold for very thin films. In fact, when $l \to 0$, E_{3t} does not tend to zero, as it should, and E_{1r} is l-independent. The neglect of the reflected wave E_{2r} in Eq. (20) allows a significant simplification of calculations for a serie of nonlinear media (Oudar, 1977; Meredith *et al.*, 1983a). In this case it is sufficient to resolve at every interface a set of two equations with two unknows (transmitted and reflected wave at this interface). A rigorous treatment of the problem consists of a simultaneous solution of a system of $2(n-1)$ equations (boundary conditions) on $2(n-1)$ variables (amplitudes of free harmonic waves).

4. Harmonic Generation in an Absorbing Medium

Equations (22)–(23) and (37)–(38) are also valid in a slightly absorbing medium at fundamental and/or harmonic frequencies. In this case the corresponding refractive indices are complex (see Chemla and Kupecek, 1971).

$$n_2^s = n_{2r}^s + i\mathscr{H}_2^s \qquad (s = \omega, 2\omega, 3\omega) \tag{40}$$

as well as the propagation angles

$$\theta_2^s = \theta_{2r}^s + i\theta_{2i}^s \tag{41}$$

For $\mathcal{H} < 1$ the real part of propagation angle is given by

$$\cos^2 \theta_{2r}^s = 1 - \frac{\sin^2 \theta_1^\omega}{(n_{2r}^s)^2} + \frac{(\mathcal{H}_2^s)^2}{2}\left[1 - \frac{\cos^2 \theta_1^\omega}{(n_{2r}^s)^2}\right] + O(\mathcal{H}^4) \tag{42}$$

for $s = \omega, 2\omega, 3\omega$, where θ_1^ω is the incidence angle.

It is seen from Eq. (42) that in a slightly absorbing medium the propagation directions of the fundamental and harmonic waves are close to those in a nonabsorbing one. The only difference will appear in the amplitude of corresponding waves. The amplitudes of fundamental and harmonic waves traversing a medium with thickness l will be given by

$$E_2^s(l) = E_2^s(0)\exp[-\mathcal{H}_2^s sl/c(\cos \theta_{2r}^s)], \qquad (s = \omega, 2\omega, 3\omega) \tag{43}$$

The wave vectors in Eqs. (22)–(23) and (37)–(38) are complex, as well as A's, α's, and corresponding transmission factors and dielectric-constant dispersion.

5. Harmonic Generation in an Infinite Medium by Focused Laser Beams

For harmonic light generation in thin nonlinear media and in weakly focusing systems, one can apply the formalism described before for plane waves. In fact, in this case the fundamental field is nearly constant inside the sample, if that is placed at the focus. This is no longer true for a thick (and/or in the case of strongly focusing systems) medium in which a large variation of fundamental field is created by focusing.

The problem of harmonic light generation by a focused laser beam has been treated by Kleinman *et al.* (1966) and by Ward and New (1969) and applied for THG in liquids by Meredith *et al.* (1983b) and in air by Kajzar and Messier (1985b) (cf. Section III.A).

For a Gaussian laser beam propagating in the z direction with beam waist $2w_0$, the third harmonic field at the point (x, y, u) is given by

$$E^{3\omega}(x, y, u) = \frac{\pi\chi^{(3)}(E_0^\omega)^3}{\Delta\varepsilon}\exp[-3(x^2 + y^2)/w_0^2(1 + i\tau)]\exp(ik^{3\omega}u)I(u) \tag{44}$$

where

$$\tau = 2(u - f)/w_0^2 k^\omega \tag{45}$$

f is the zth coordinate of focus, and $I(u)$ is a sum of all contributions to the harmonic field from $-\infty$ to $z = u$,

$$I(u) = \frac{3\omega}{c} \Delta n \int_{-\infty}^{u} dz' \frac{e^{i\phi}}{(1 + i\tau')^2} = Ce^{i\psi} \tag{46}$$

with

$$\phi = \frac{3\omega}{c}(z' - u)\,\Delta n \tag{47}$$

and

$$\tau = 2(z' - f)/w_0^2 k^\omega \tag{48}$$

The modulus C of the function $I(u)$ is generally smaller than 1 and tends to 1 when $u - f$ and $w_0 \to \infty$. For $u \to \infty$, $I(u) \to 0$. This means that there is no resultant harmonic field in an infinite medium. By placing a sample in focus one introduces a phase mismatch between harmonic fields created before and after the sample, which will now interfere with harmonic field created in the sample itself. It shows importance of environmental effects for samples placed in air, discussed in Section III.A.

6. Electric-Field-Induced Second-Harmonic Generation

For electric-field-induced second-harmonic generation in a finite-thickness nonlinear medium immersed between two linear media, Eqs. (22) and (23) apply directly by changing 3ω to 2ω. The corresponding coherence lengths are

$$l_c = |\lambda_\omega/4(n_2^\omega - n_2^{2\omega})| \qquad \text{and} \qquad l_c^r = |\lambda_\omega/4(n_2^\omega + n_2^{2\omega})|$$

B. Graphical Representation

It is sometimes convenient to represent the resulting harmonic field generated in a serie of nonlinear media by

$$E^{3\omega} = \sum_j T_j E_j^{3\omega} \tag{49}$$

where the T_j are whole transmission factors and $E_j^{3\omega}$ is the harmonic field generated in medium j, in a complex plane. According to Eq. (22) (neglecting the second term with A_2) in a nonabsorbing medium, the harmonic field E_{3t} will be represented by AM (see Fig. 4) with points A and M lying on a circle with radius $R = A_0 A_1$. The angle $\Delta\psi = 3\omega(N_2^\omega - N_2^{3\omega})\,l/c$ is the phase mismatch between the bound and free wave in this medium. For a serie of nonlinear media, the harmonic fields will be represented by similar vectors lying on circles with radii $R_j = (A_0 A_1 T)_j$ (cf. Fig. 4) with corresponding angles $\Delta\psi_j$. The real and imaginary parts of the resultant field will be represented by the sum of projections of these vectors on the x and y axes, respectively.

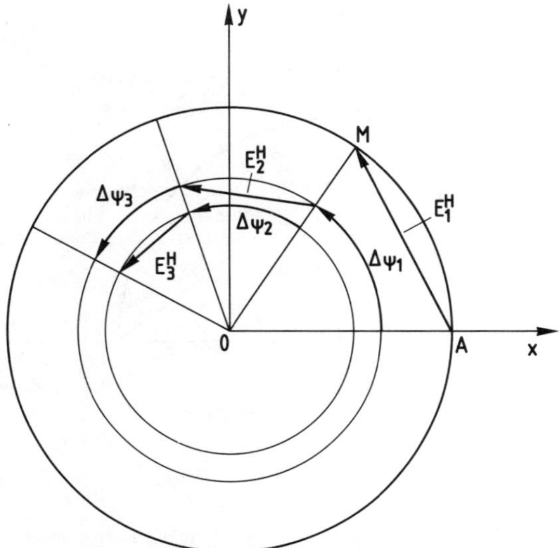

Fig. 4. Graphical representation of harmonic field summation in three successive nonlinear media, without absorption, in complex plane. The real and imaginary parts of resultant harmonic field are sums of corresponding real and imaginary parts of harmonic fields created in successive media. For a complex $\chi^{(3)}$ susceptibility with phase ϕ the corresponding vector should be rotated by angle ϕ. In an absorbing medium, points A and M do not lie on the same circle but on a spiral with smaller radius.

For a focused laser beam in an infinite medium, the harmonic field increases progressively when going from $-\infty$ to $z = u$, and the representative point M will describe a progressively growing spiral, the resultant harmonic field being equal to the length of 0M (see Fig. 5).

In an absorbing medium, points A and M (see Fig. 4) do not lie on the same circle but on circles with different radii:

$$0A' = 0A \exp(-3\mathcal{H}_2^{3\omega}\omega l/c \cos \theta_2^{3\omega}) \tag{50}$$

$$0M' = 0M \exp(-3\mathcal{H}_2^{\omega}\omega l/c \cos \theta_2^{\omega}) \tag{51}$$

In the case of the $\chi^{(3)}$ complex ($\chi^3 = |\chi^3|e^{i\phi}$), the vector AM (or A'M') has to be rotated by an angle ϕ.

III. EXPERIMENTAL METHODS AND RESULTS

A. Experimental Conditions and Environmental Effects

A typical experimental set-up for THG and EFISHG measurements is shown in Fig. 6. The fundamental laser beam yielded by a Q-switched

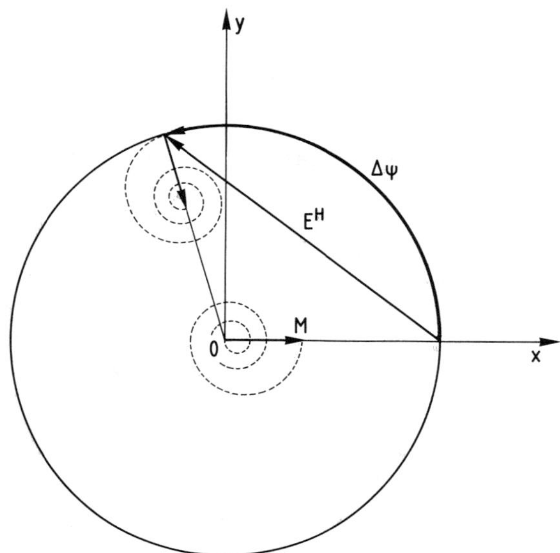

Fig. 5. Graphical representation of harmonic field summation in the case of a nonlinear medium lying between two infinite nonlinear media for focused laser beams.

Nd: YAG laser operating at 1.064 μm and converted to 0.532 μm by a phase-matched KDP single crystal is pumping a dye laser. The outgoing dye laser beam with width of 0.12 cm^{-1} is subsequently shifted in a high-pressure ($\sim$56 bars) hydrogen cell by 4155 cm^{-1} or its multiple through a stimulated

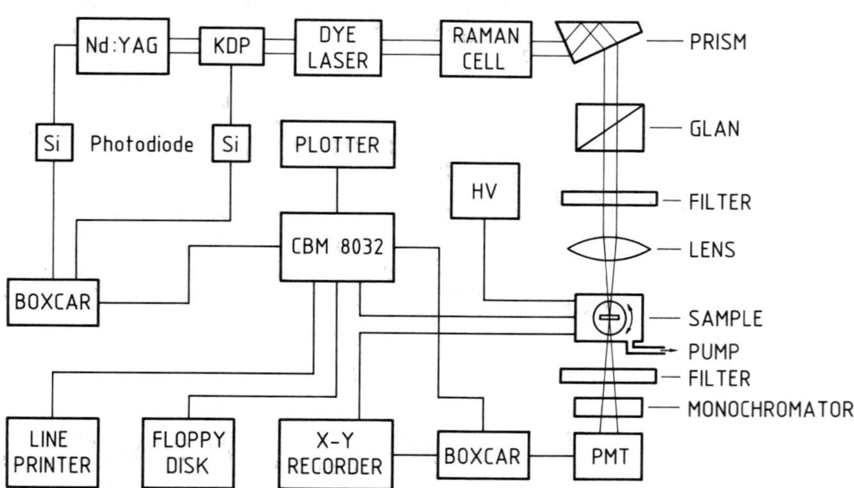

Fig. 6. Experimental setup for EFISHG and THG measurements.

Raman scattering process. This system allows the laser wavelength variation in the range 0.53–1.907 μm. The dispersive prism permits a separation of other Raman frequencies and the fundamental one. A system of filters allows elimination of the parasitic light. The laser beam is focused on a sample mounted on a goniometer, which can be translated and/or rotated by stepping motors controlled by a minicomputer. The sample assembly is placed in a vacuum cell with fused-silica windows. The vacuum cell length can be adjusted in order to avoid window effects. The harmonic light signal generated in the sample, selected by a Jobin-Yvon monochromator, and detected by a photomultiplier, integrated by a box car, is stored on floppy disk as a function of incidence angle and/or translation. The fundamental λ_ω and/or its doubled $\lambda_{2\omega}$ wavelength fluctuations are monitored by a rapid Si photodiode, which again is read by minicomputer through a box-car integrator. This computer-controlled system allows precise and repetitive determination of harmonic intensities as a function of sample position. Such a possibility is important for the $\chi^{(3)}$ phase determination, as will be discussed later.

The use of the vacuum cell is very important if one wants to avoid environmental effects in THG measurements. In fact, as was first observed by Meredith *et al.* (1983a), the laser beam also generates a harmonic field in surrounding sample air that interferes with that created in the sample itself, modifying the resulting intensities. Figure 7 shows the harmonic-light

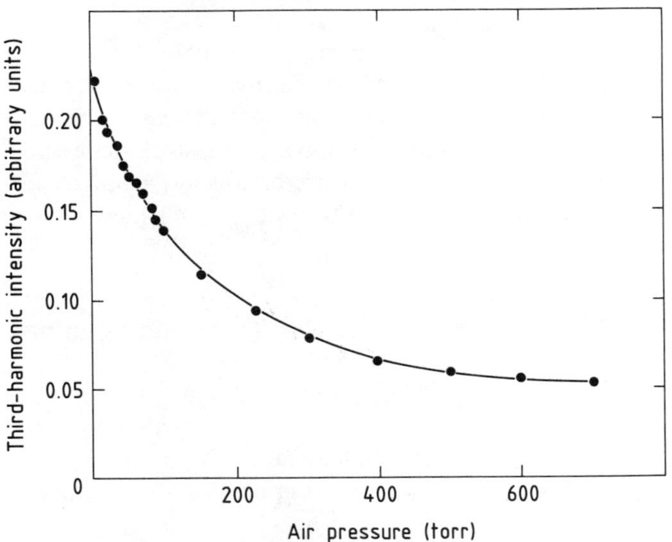

Fig. 7. Maximum third-harmonic intensity from a silica plate as a function of air pressure at 1.064 μm fundamental wavelength and focal length of 50 cm. [After Kajzar and Messier (1985b).]

intensity dependence on air pressure for a focal length of 500 mm. A factor of about four in intensity decrease between vacuum and ambient pressure is observed.

The influence of surrounding air on harmonic-light intensities has been studied in detail by Kajzar and Messier (1985b). These authors proposed also an experimental correction, based on harmonic generation by the focused beams formalism described in Section II that takes it into account. For the simplest case of a slab in air, this corresponds to considering three nonlinear media: two infinite and one finite (see Fig. 5). Thus the resulting harmonic intensity will follow the formula.

$$J^{3\omega} \propto \left| E/[\chi^{(3)}/\Delta\varepsilon]_A + \frac{C'(n^{\omega} + n^{3\omega})_G}{2}[T^{3\omega}e^{i(\psi+\theta_1)} - (T^{\omega})^3 e^{-i(\psi+\theta_2)}](E^{\omega})^3 \right|^2 \tag{52}$$

where E is a resulting harmonic field created in the slab, C' is the air contribution parameter $[C' = C(\chi^{(3)}/\Delta n)_A/(\chi^{(3)}/\Delta n)_G]$ with A air and G glass, ψ is the phase of air contribution [see Eq. (46)], θ_1 and θ_2 are phases of harmonic and fundamental waves relative to a reference one, T^{ω} and $T^{3\omega}$ are transmission factors of fundamental and harmonic light ($T^{\omega} = t_{12}^{\omega}t_{23}^{\omega}$ and $T^{3\omega} = t_{23}^{3\omega}$), and E^{ω} is the incident field.

The magnitude of the air contribution parameter C' depends on the focal length, laser wavelength, and silica-plate thickness. For a given wavelength, the value of C' increases with focal length and decreases with silica-plate thickness.

For a silica plate with 1 mm thickness, the air contribution modifies the envelope of Maker fringes. For $C' < 0.44$, the envelope decreases more rapidly with incidence angle than in a vacuum, whereas it increases for $C' > 0.44$ (see Fig. 8). For $C' \approx 0.44$ there is a cancellation between silica-plate signal and air contribution. The interference spectra are completely changed for a filled liquid cell in the case of large air contribution. In all these cases, Eq. (52) permits us to describe these behaviors.

B. THG by Transmission in PDA Thin Films and Solutions

For the sake of simplicity, we neglect hereafter environmental effects, and the formulas given below are valid for THG measurements in a vacuum. In the case when the environmental effects intervene, the harmonic intensity should be corrected for air contribution as described in the preceding section [Eq. (52)].

First we consider THG from a thin polymer film. The large values of cubic susceptibility allows THG measurements from even very thin films [a few

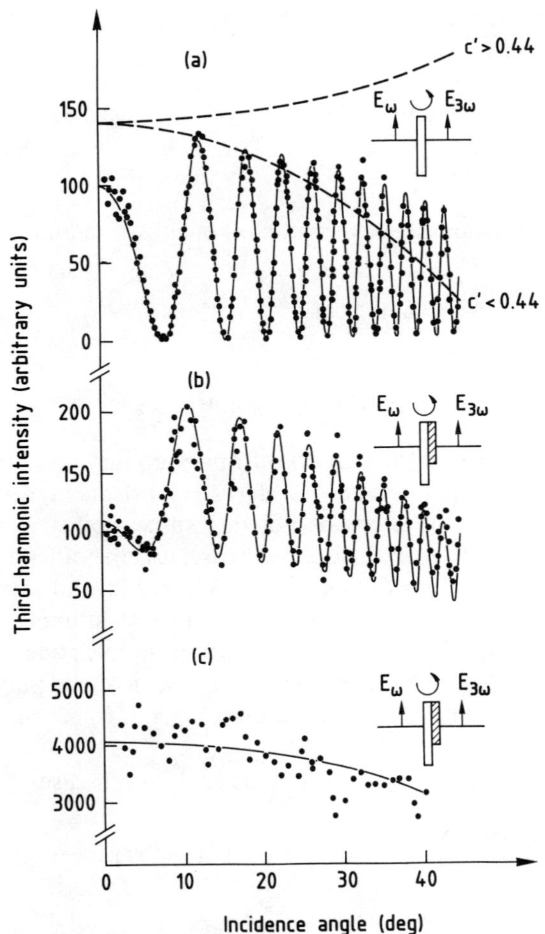

Fig. 8.　Maker fringes from (a) a silica plate (b) silica plate with a polysilane, and (c) LB film on one side of substrate, measured in vacuum. Broken lines in (a) represent envelope variation of third-harmonic intensities from a silica plate in the case of small ($C' < 0.44$) and large ($C' > 0.44$) air contributions.

hundred angstroms (Kajzar *et al.*, 1983b)]. The most adequate microstructures are Langmuir–Blodgett thin films, described elsewhere (Barraud and Vandevyver, Chapter III-5). These films obtained from monomers have a well-defined thickness and a high degree of organization in the direction perpendicular to the substrate plane (see, e.g., Lieser *et al.* 1960; Kajzar and Messier, 1981, 1983). Two experimental configurations for THG measurements are possible.

1.   Thin film on one side of the substrate. In this case, the third-harmonic intensity is given by

$$J^{3\omega} = \frac{64\pi^4}{c^2}\left|e^{i(\psi_F^S + \psi_H^P)}(\chi^{(3)}/\Delta\varepsilon)_s[T_1(1 - e^{-i\Delta\psi_s}) + \rho e^{i\phi}(e^{i\Delta\psi_P} - 1)T_2]\right|^2 (J^\omega)^3$$

(53)

where $\Delta\psi_S$ and $\Delta\psi_P$ are phase mismatches in substrate and in polymer film, respectively, and

$$\Delta\psi = 6\pi(n^\omega \cos\theta^\omega - n^{3\omega}\cos\theta^{3\omega})l/\lambda_\omega$$

(54)

and

$$\rho = (|\chi^{(3)}|/\Delta\varepsilon)_P/(\chi^{(3)}/\Delta\varepsilon)_S$$

(55)

The T_1 and T_2 factors in Eq. (53) arise from boundary conditions and from transmission. Subscripts S and P refer to substrate and polymer film, respectively, and ϕ is the phase of polymer susceptibility provided that of substrate is equal to zero (in the opposite case, it is the difference).

In cases where $l_P \ll l_c^P$, $\Delta\psi_P$ is very small ($\Delta\psi_P \ll \pi/2$) and varies slowly with incident angle. The resulting harmonic intensities as a function of incidence angle are a monotonic function for $\rho \gg l_c^S/l_P$ or a complex one for $\rho \leqslant l_c^S/l_P$. In the last case it is possible to determine the phase ϕ of the polymer if all the characteristics of substrate are known (e.g., from an independent measurement).

2.   Thin films on both faces of the substrate. In this case,

$$J^{3\omega} = \frac{64\pi^4}{c^2}|(\chi^{(3)}/\Delta\varepsilon)_S e^{i\psi}[\rho T_1 e^{i\phi}(1 - e^{-i\Delta\psi_P})$$

$$+ T_2(e^{i\Delta\psi_S} - 1) + \rho T_3 e^{i(\Delta\psi_S + \phi)}(e^{i\Delta\psi_P} - 1)]|^2(J^\omega)^3$$

(56)

where $\psi = \psi_F^P + \psi_H^P + \psi_H^S$ and the T_i are factors arising from transmission and boundary conditions. Subscripts F and H refer to fundamental and harmonic wave, respectively. In Eq. (56) we have assumed that both polymer films have the same thickness, as provided in the Langmuir–Blodgett (LB) technique.

Similarly, $l_P \ll l_c^P$ and Eq. (56) lead always to a complex interference pattern. In a special case, $\rho \gg l_c^S/l_P$, the interference patterns are similar to the Maker fringes (see Fig. 8), with the only difference being in their origin. In the case of Maker fringes, the interference maxima and minima originate from constructive or destructive interference between free and bound waves, whereas in the present case they are due to interference between free harmonic waves generated in films on both sides of substrate; the phase mismatch is mainly due to the refractive index dispersion in the substrate itself ($\Delta\psi_S$).

First THG measurements on PDA/LB films performed by Kajzar et $al.$ (1983a) at 1.064 μm and calibrated with a phase-matched THG from a CaCO$_3$ single crystal gave a $\chi^{(3)}$ value of about 2 orders of magnitude smaller than that reported by Sauteret et $al.$ (1976) from THG measurements on a PTS single crystal [$\chi^{(3)} = (8.5 \pm 5) \times 10^{-10}$ esu]. Our recent phase-dispersed THG measurements (Kajzar and Messier, 1986) on a "blue" form of PDA LB film, $\mathrm{R} = (\mathrm{CH_2})_8\mathrm{COO\,Cd}^{1/2}$, $\mathrm{R'} = (\mathrm{CH_2})_{15}\mathrm{CH_3}$, calibrated with THG from a silica plate allows us to elucidate this discrepancy. As it is seen in Fig. 9, there exist two resonance enhancements in the $\chi^{(3)}_{xxxx}(-3\omega;\omega,\omega,\omega)$ value in the fundamental wavelength region 0.8–1.9 μm. The resonant value at 1.907 μm, corrected for polymer disorder and equal to $(2.2 \pm 0.2) \times 10^{-10}$ esu, is close to

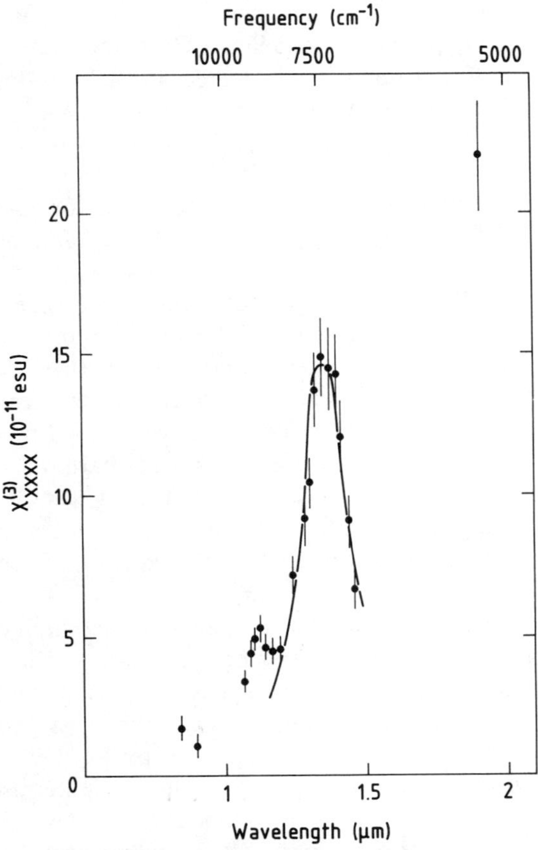

Fig. 9. Cubic susceptibility $\chi^{(3)}_{xxxx}(-3\omega;\omega,\omega,\omega)$ of the blue form of polydiacetylene LB film, corrected for polymer disorder, as a function of incident laser wavelength. Solid line represents a fit of Eq. (10). [After Kajzar and Messier (1986).]

that measured by Sauteret *et al.* (1976) at the same wavelength. It is seen from Eq. (10) that this enhancement is due to a three-photon resonance, the triple of the incident photon energy being equal to the first excitonic transition energy. The second resonance at 1.35 μm is a two-photon resonance. The two-photon level with width $\Gamma = 1160$ cm^{-1} lies 14,800 cm^{-1} above the fundamental state and about 800 cm^{-1} below the first optically allowed level.

Similarly, cubic susceptibilities and microscopic hyperpolarizabilities can be also obtained from THG in polymer solutions. This technique also allows determination of electronic hyperpolarizabilities far from vibronic transitions.

Several techniques have been proposed for THG measurements in liquids. The simplest one, proposed by Meredith *et al.* (1983b), in which the harmonic light is generated in a thick liquid cell (6 cm) by focused laser beam, uses a formalism described for an infinite medium. In view of the large optical pathlength, this technique can be used only in the case of liquids transparent at both fundamental and harmonic frequencies. Other techniques use a principle of superposition of harmonic field created in cell windows and liquid compartment. The resultant harmonic field is then given by

$$E^{3\omega} = T_1 E_{G1}^{3\omega} + T_L E_L^{3\omega} + T_2 E_{G2}^{3\omega} \tag{57}$$

where the T's are corresponding whole transmission factors and the subscripts G1, L, and G2 refer to input, liquid compartment, and output window, respectively.

Meredith *et al.* (1983a) use a wedged liquid cell with prismed windows. The harmonic light variation is achieved by the cell translation. Thalhammer and Penzkofer (1983) use also a wedged liquid cell, but the input and output window have thickness equal to an even number of coherence lengths. In this case they obtain zeroth harmonic fields in both windows ($E_{G1}^{3\omega} = E_{G2}^{3\omega} = 0$) and, similar to Meredith *et al.* (1983a), they obtain the harmonic field variation through liquid cell translation perpendicular to the laser beam. This technique is simple, but the use of the liquid cell is limited to a given wavelength and it also requires high-quality window surfaces and high precision in their fabrication (thickness).

Kajzar and Messier (1985a) use a plane-parallel liquid cell. Optical pathlength variation is realized by sample rotation around an axis perpendicular to the laser beam. The technique can be used for slightly absorbing liquids by diminution of liquid compartment thickness. In this case the interference between the harmonic fields given by Eq. (57) gives rise to a complex fringes picture that can be resolved with computer use (see Fig. 10). The technique gives uniquely not only the modulus of $\chi^{(3)}$ but also its phase. It permits us to avoid multiple reflection effects, which may be important at normal incidence. The harmonic field of the studied liquid is directly calibrated with that of the cell windows.

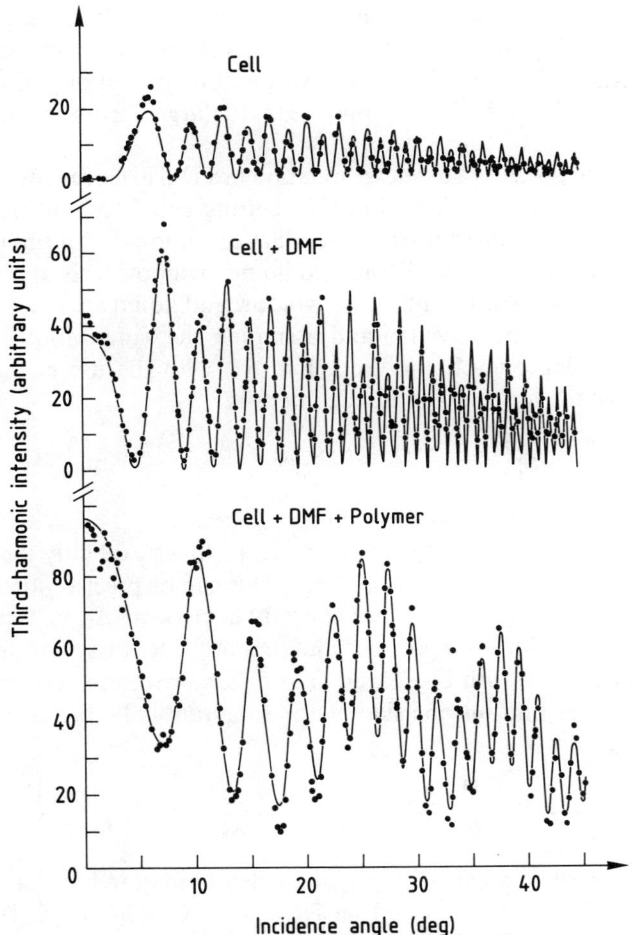

Fig. 10. Successive harmonic interference patterns from a liquid cell with liquid compartment thickness of 200 μm as a function of its filling at 1.064 μm. Because of polymer absorption at harmonic frequency, the input-window harmonic emission is completely absorbed by polymer solution. [From Kajzar and Messier (1985a).]

C. Determination of the Modulus of $\chi^{(3)}(-3\omega;\, \omega,\, \omega,\, \omega)$ and Its Phase

The interference of harmonic fields generated in the substrate and in the polymer film or in the cell window and in the liquid allows us to determine the modulus and the phase of cubic susceptibility $\chi^{(3)}(-3\omega;\omega,\omega,\omega)$ of the polymer itself, provided that those of the reference (substrate or cell windows) are known. As will be discussed in the next section, the knowledge of phase gives important information on the position and parity of excited states and is useful for practical applications (e.g., two-photon absorptions). On the other hand, neglect of phase in $\chi^{(3)}$ may lead to large errors in its modulus determination.

The possibility of $\chi^{(3)}$ phase determination in THG experiments on polymer solutions follows directly from Eq. (57). Setting equal to 1 all transmission factors as well as those arising from boundary conditions [A_1 terms in Eq. (22)] on the interfaces of window–liquid and liquid–window (this approximation holds well only if refractive indices of window and liquid are close, as it is the case for some organic solvents) and assuming the same input and output window thickness, one obtains from Eq. (57) with the use of Eq. (22) the following formula for third-harmonic intensity:

$$J^{3\omega} = \frac{256\pi^4}{c^2}\, T \left| \frac{\chi_L^{(3)}}{\Delta\varepsilon_L} \sin\frac{\Delta\psi_L}{2} + \frac{2\chi_G^{(3)}}{\Delta\varepsilon_G} \cos\frac{\Delta\psi_L + \Delta\psi_G}{2} \sin\frac{\Delta\psi_G}{2} \right|^2 (J^\omega)^3 \quad (58)$$

where T is a whole transmission factor.

Although Eq. (58) is an approximate one, it shows explicitly the harmonic intensity dependence on the phase of $\chi_L^{(3)}$. This can be precisely determined if $\chi_G^{(3)}$ is known as well as the dielectric-constant dispersions $\Delta\varepsilon_{L,G}$. Equation (58) holds also for THG from a polymer film on both sides of substrate inverting $\chi^{(3)}$ terms in Eq. (58) with L → S and G → P (S, substrate; P, polymer).

For a polymer film on one side of the substrate only, the corresponding formula is given by

$$J^{3\omega} = \frac{256\pi^4}{c^2}\, T \left| \frac{\chi_S^{(3)}}{\Delta\varepsilon_S} e^{-i\Delta\psi_S/2} \sin\frac{\Delta\psi_S}{2} + \frac{\chi_P^{(3)}}{\Delta\varepsilon_P} e^{i\Delta\psi_P/2} \sin\frac{\Delta\psi_P}{2} \right|^2 (J^\omega)^3 \quad (59)$$

The THG measurements with $\chi^{(3)}$ phase determination have been done on solutions of PTS-12 polydiacetylene [R = R' = $(CH_2)_4OSO_2C_6H_4CH_3$] in dimethylformamide (DMF) at 1.064 and 1.907 μm. The used liquid cell was made from Suprasil I, with wide transparency range (0.18–2.5 μm). At 1.907 μm the measured cubic susceptibility of solution is a linear function of polymer concentration, as it should be for $\chi^{(3)}$ being real and positive (Kajzar and Messier, 1985a). At this wavelength, both the fundamental and the harmonic frequency lie at the polymer transparency range, below the optical gap.

At 1.064 μm the fundamental frequency lies at the transparency range, whereas the harmonic one lies above the allowed excitonic state. In this case $\chi^{(3)}(-3\omega; \omega, \omega, \omega)$ is complex with negative real and imaginary parts (see Fig. 11). The detailed analysis of this resonance allows us to place the two-photon state (with even parity) above 18,900 cm^{-1} (0.53 μm).

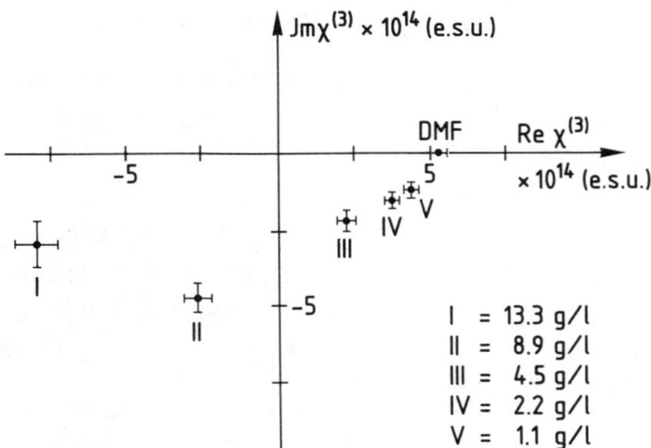

Fig. 11. Concentration variation of real and imaginary parts of cubic susceptibility $\chi^{(3)}(-3\omega; \omega, \omega, \omega)$ of a PTS-12 solution in dimethylformamide. [From Kajzar and Messier (1985a).]

We note here that special care has to be paid when a $\chi^{(3)}$ phase determination is performed. Generally the harmonic frequency (or fundamental one) lies at the absorption region (one- and two-photon levels are often very close), and refractive indices are complex. Similarly complex are transmission factors and those arising from boundary conditions, as well as dielectric-constant dispersion. All these factors introduce an additional phase, which can be confused with the phase of $\chi^{(3)}$ if these are not taken into account.

D. EFISHG in Thin Films and in Solutions

As we mentioned before, polydiacetylenes have a centrosymmetric structure and no SHG is observed. This symmetry of polymers can be broken by applying an external dc field, and a net dipolar moment can be induced in its direction. In this case an SHG will be observed and the induced nonlinear polarization [see Eq. (14)] will be proportional to the external field strength E_0.

The EFISHG experiments (Chollet et $al.$, 1985a) have been done by transmission on PDA LB multilayers [$R = (CH_2)_8COOCd^{1/2}$, $R' = (CH_2)_{13}CH_3$] with thickness $l_P = 0.05-0.2$ μm and on thin films prepared from a soluble polymer [3-BCMU, $R = R' = (CH_2)_3OCONHCH_2COOC_4H_9$] by solvent evaporation or by dipping technique ($l_P = 0.2-10$ μm). Whereas the LB films are well crystallized with a preferential orientation of polymer chains parallel to the substrate, those of 3-BCMU are partially amorphous.

1. Frequency Variation of $\chi^{(3)}(-2\omega;\omega,\omega,0)$

The wave-dispersed EFISHG measurements were performed on the blue form of LB thin film in the fundamental wavelength range $0.8-1.4\ \mu m$. The SH intensities have been calibrated with a wedged α-quartz single crystal. There exists a simple relation between maximum SH intensities $J_Q^{2\omega}$ from α-quartz and EFISH intensity from polymer film (due to the small polymer film thickness l_P, the SH intensity is directly proportional to l_P^2), and neglecting the harmonic field generated in substrate itself [this is negligible compared to that created in polymer, because of large value of $\chi^{(3)}(-2\omega;\omega,\omega,0)$ for polydiacetylenes],

$$\frac{\chi^{(3)}(-2\omega;\omega,\omega,0)E_0}{\chi_Q^{(2)}(-2\omega;\omega,\omega)} = \frac{8}{3}\frac{l_c^Q}{\lambda\omega}\frac{F_P}{(n^\omega + n^{2\omega})_Q}\left|\frac{T_Q}{T_P}\right|\left(\frac{J_P^{2\omega}}{J_Q^{2\omega}}\right)^{1/2} \tag{60}$$

where subscripts Q and P refer to quartz and polymer film, respectively; T_Q and T_P are whole transmission factors; and

$$F_P = \left|\frac{(n^\omega)^2 - (n_r^{2\omega} + i\mathcal{H}^{2\omega})^2}{e^{i\alpha(n^\omega - \mathcal{H}^{2\omega})} - e^{-i\alpha\mathcal{H}^{2\omega}}}\right|_P \tag{61}$$

where $\alpha = 2\pi l_P/\lambda_\omega$. The factor F_P takes into account the polymer absorption at the harmonic frequency [for nonabsorbing film, $F_P \approx (n^\omega + n^{2\omega})_P\lambda_\omega/l_P)$]. For the imaginary part of the refractive index, $\mathcal{H}_P^{2\omega}$ satisfying the relation

$$|n^\omega - n_r^{2\omega}|_P \ll \mathcal{H}_P^{2\omega} < (n^\omega + n_r^{2\omega})_P \tag{62}$$

the factor F_P is given by

$$F_P = (n^\omega + n_r^{2\omega})_P\mathcal{H}_P^{2\omega}/[1 - \exp(-\alpha\mathcal{H}_P^{2\omega})] \tag{63}$$

Corrected for the polymer chain disorder, values of $\chi_{xxxx}^{(3)}(-2\omega;\omega,\omega,0)$ are plotted in Fig. 12 as a function of the fundamental laser wavelength. As in the THG experiment, a maximum in $\chi_{xxxx}^{(3)}(-2\omega;\omega,\omega,0)$ is observed at around $1.35\ \mu m$.

The fact that one observes smaller values of $\chi_{xxxx}^{(3)}(-2\omega;\omega,\omega,0)$ compared to those of the $\chi_{xxxx}^{(3)}(-3\omega;\omega,\omega,\omega)$ tensor component is probably due to different morphology and degree of polymerization of thin films used in both experiments, as well as different local field factors.

2. EFISHG in Polymer Solutions

EFISHG measurements have been done on PTS-12 [$R = R' = (CH_2)_4$-$OSO_2C_6H_4CH_3$] solutions in DMF and in $CHCl_3$ at $1.064\ \mu m$ using a liquid cell described by Oudar (1977). This technique allows also a determination of the modulus of $\chi^{(3)}(-2\omega;\omega,\omega,0)$ susceptibility and of its phase (Kajzar *et al.*, 1986).

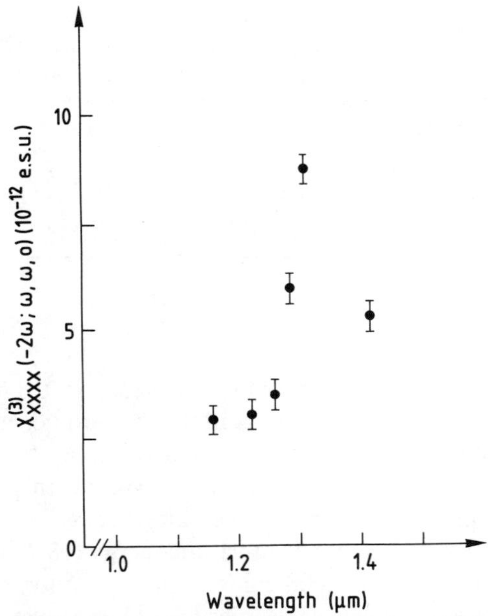

Fig. 12. Cubic susceptibility $\chi^{(3)}(-2\omega; \omega, \omega, 0)$ corrected for polymer disorder of blue form of polydiacetylene LB film as a function of incident laser wavelength. [From Chollet *et al.* (1986).]

As in THG experiments on the same polymer solutions, one observes a complex value for this susceptibility tensor at 1.064 μm with a negative real (the sign of the imaginary part is undetermined) part. The values of $\chi^{(3)}(-2\omega; \omega, \omega, 0)$ and molecular hyperpolarizabilities γ per monomer unit are give in Table I, together with other determinations.

3. Polarization Effects in PDA Thin Films

The dc field in Eq. (14) may be reduced in a dielectric by internal polarization directed opposite to the applied external field. In this case the molecules are treated with an effective field $E_e = E_0 - E_p$, where E_p is the internal polarization field. In the case when E_p is equal to E_0, no SHG will be observed. Such a disappearance of SH intensity is observed in PDA thin films at 1.064 μm. In Fig. 13, the observed SH intensity from an LB thin film is plotted as a function of illumination time. A rapid initial decrease of $J^{2\omega}$ is observed. This phenomenon is due to an internal polarization created in a zone illuminated by the laser beam and cancelling the applied external field. When the external field is switched off, the SH signal reappears due to the polarization field E_p. The polarization field E_p is largest for LB films, where it is equal to the applied external field (5×10^4 V/cm), and smaller in

TABLE I

Cubic Susceptibility and Molecular Hyperpolarizability per Monomer Unit of Soluble PTS-12 [R = R′ = $(CH_2)_4OSO_2C_6H_4CH_3$] Polydiacetylene at 1.064 μm

Solvent	Re $\chi^{(3)} \times 10^{11}$ esu	Im $\chi^{(3)} \times 10^{11}$ esu	Re $\gamma \times 10^{33}$ esu	Im $\gamma \times 10^{33}$ esu
DMF[a]	−0.8	−0.5	−0.36	−0.22
DMF[b]	−0.40	±0.53	−0.20	±0.26
$CHCl_3$[b]	−0.37	±0.08	−0.18	±0.04

[a] THG (Kajzar and Messier, 1985a).
[b] EFISHG (Kajzar et al., 1986).

semicrystalline 3-BCMU films (Chollet et al., 1985a). In LB films the internal polarization remains a few tens of hours in the dark and disappears in visible or 1.064-μm laser-light illumination. This polarization is observed for incident laser wavelength smaller than 1.1 μm in the polymer transparency region. It means that the creation of electron–hole pairs goes through a two-photon process and that the band-to-band transition energy (18,900 cm^{-1}) is higher than the optical-gap energy (15,600 cm^{-1}). The separation of electron–hole pairs in an external electric field takes place over larger distances in LB than

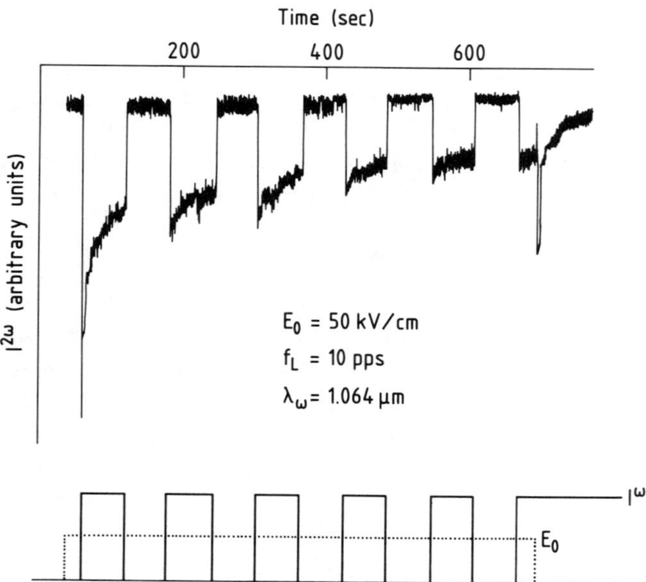

Fig. 13. Temporal second-harmonic intensity variation in function of laser illumination and external dc electric field application from an LB thin film. A dc field switch-off results in SH signal increase due to internal polarization. [From Chollet et al. (1986).]

in 3-BCMU films. This is probably connected with structure and film morphology.

From the value of the two-photon absorption coefficient given by Lequime and Hermann (1977) [see also Shand and Chance (1978)] one can calculate the mean polarization per unit volume, which is given by

$$P = N\sigma d \tag{64}$$

where N is the number of created charge carriers, σ is their separation probability, and d is the average distance of electron–hole pair separation. For an incident laser beam intensity of 150 MW/cm^2 focused on a surface with $D = 200\ \mu$m diameter during 13 nsec one obtains $N = 5 \times 10^{19}$ e/cm^3. On the other hand, the polarization field $E_p = 5 \times 10^4$ V/cm leads in a dielectric with $\varepsilon = 2.5$ to a polarization $P = 7 \times 10^{10}$ e/cm^2. It means that both probability separation and average distance of separation are very small ($\sigma \ll 1$ and $d \ll D$).

E. Four-Wave Mixing Experiments in Polymer Solutions

Using two incident laser beams with frequency ω_1 and ω_2 and measuring mixed-frequency generation, Chance $et\ al.$, (1980) studied $\chi^{(3)}(-2\omega_1 + \omega_2; \omega_1, \omega_1, \omega_2)$ susceptibility as a function of ω_1 in the visible range. The measurements were done on yellow, red, and blue solutions (the solution color depending on optical energy gap value E_{ng}, cf. Fig. 1) of 3-BCMU and 4-BCMU (R $=$ R$' = $ (CH$_2$)$_4$ OCONHCH$_2$-COOC$_4$H$_9$) in function of polymer concentration. From the concentration dependence of $\chi^{(3)}(-2\omega_1 + \omega_2; \omega_1, \omega_1, -\omega_2)$, Chance $et\ al.$ (1980) determined its modulus and phase for these polymers and consequently the positions and widths of two photon levels. The results obtained by these authors are listen in Table II. In all cases

TABLE II

Gap (E_{ng}) and Two-Photon State (E_{mg}) Energies in Polymer
Solutions and in Blue-Form LB Film with Two-Photon
State Width Γ

Polymer form	E_{ng} (cm^{-1})	E_{mg} (cm^{-1})	Γ (cm^{-1})
Yellow solution[a]	21,300	30,500	4600
Red solution[a]	18,900	28,100	4000
Blue solution[a]	15,900	23,200 20,500	2200
Blue LB film[b]	15,600	14,800	1160

[a] Chance $et\ al.$ (1980).
[b] Kajzar and Messier (1986).

they found the two-photon level lying above the first optically allowed state. The position of the two-photon state depends on the gap energy E_{ng} and approaches the one-photon state when E_{ng} decreases (passing from yellow to blue solution). The width of the two-photon state (see Table II) is large and narrows when E_{ng} decreases.

IV. DISCUSSION

A. Delocalization and Resonance Effects

1. *Three-Photon Resonance*

As was discussed in the introduction, one expects a strong dependence of cubic susceptibility of one-dimensional conjugated systems on delocalization length. In fact, the THG measurements of Sauteret *et al.* (1976) at 1.907 μm show a three order of magnitude increase in $\chi^{(3)}_{xxxx}(-3\omega; \omega, \omega, \omega)$ susceptibility between monomer and polymer. We know now that the value reported by Sauteret *et al.* is a three-photon resonant value. In fact, when the triple of the incident photon energy approaches to the gap energy, the cubic suceptibility $\chi^{(3)}(-3\omega; \omega, \omega, \omega)$ will be enhanced. In this case, conserving only dominant term in Eq. (10), one obtains for real and imaginary parts of cubic susceptibility in a three-level approximation ($n = n'$) the expressions

$$\mathrm{Re}\, \chi^{(3)}(-3\omega; \omega, \omega, \omega) \propto \frac{E_{ng} - 3\omega}{[(E_{ng} - 3\omega)^2 + \Gamma^2](E_{ng} - \omega)(E_{mg} - 2\omega)} \quad (65)$$

and

$$\mathrm{Im}\, \chi^{(3)}(-3\omega; \omega, \omega, \omega) \propto \frac{\Gamma}{[(E_{ng} - 3\omega)^2 + \Gamma^2](E_{ng} - \omega)(E_{mg} - 2\omega)} \quad (66)$$

At 1.907 μm the triple of the photon energy for the blue form of polydiacetylene is close to the optical gap energy ($E_{ng} \approx 15{,}600\ cm^{-1}$), and a resonant enhancement in $\chi^{(3)}$ is observed.

Equations (65) and (66) show also that at resonance the $\chi^{(3)}$ susceptibility is complex with no real part.

2. *Two-Photon Resonance*

Besides the three-photon resonance at 1.907 μm, the THG measurements show the existence of a second enhancement in $\chi^{(3)}(-3\omega; \omega, \omega, \omega)$ at 1.35 μm (see Fig. 9). Due to the facts that the fundamental laser frequency lies at the polymer transparency region and the harmonic frequency in the optical absorption deep, the only explanation of this resonance is via a two-photon process. A possibility of such resonance enhancement is given by Eq. (10) for the double of fundamental photon energy equal to the energy difference between m (with even parity) and fundamental levels. Again, in a three-level approximation and conserving only dominant terms, one gets for cubic susceptibility

$$\chi^{(3)}(-3\omega; \omega, \omega, \omega) \propto \frac{1}{(E_{mg} - 2\omega - i\Gamma)(E_{ng} - \omega)(E_{ng} - 3\omega)} \tag{67}$$

As for a three-photon resonance, the real part of $\chi^{(3)}$ disappears at resonant frequency $(2\omega = E_{mg})$.

The determined position and width of the two-photon level are given in Table II. For the blue form of LB film of polydiacetylene, the two-photon level $(E_{mg} = 14,800 \text{ cm}^{-1})$ with width $\Gamma = 1160 \text{ cm}^{-1}$ lies below the first optically allowed transition $(E_{ng} = 15,600 \text{ cm}^{-1})$. A similar situation was also observed in polyenes, where the two-photon state lies also below the one-photon level (Hudson and Kohler, 1974). This result is in contradiction with that of Chance *et al.* (1980), who found the two-photon state above the one-photon level (see Table II). These authors observe, however, that this level approaches the one-photon state passing from yellow to blue polymer solution (decreasing E_{ng}). This means that the position of the two-photon state may depend strongly on the environment.

The existence of a two-photon resonance at 1.35 μm is also confirmed by EFISHG measurements on the blue form of the polymer (see Fig. 11). For a two-photon resonance, the dominant terms in corresponding cubic susceptibility have the following form (Ward, 1965):

$$\chi^{(3)}(-2\omega; \omega, \omega, 0) \propto \frac{1}{(E_{mg} - 2\omega - i\Gamma)(E_{ng} - \omega)} \sum_{n'} \left(\frac{1}{E_{n'g} - 2\omega} + \frac{1}{E_{n'g}} \right) \tag{68}$$

Although Eq. (68) is resonant with an odd parity level (n') for two-photon transitions, the position of resonance with respect to the polymer optical absorption spectrum indicates that it is also a resonance with even parity level.

We note here that THG (Kajzar and Messier, 1985a) and EFISHG (Kajzar *et al.*, 1986) measurements on polymer solutions with corresponding $\chi^{(3)}$ susceptibility phase determination at 1.064 μm allow us to place the two-photon level above 0.53 μm (18,900 cm^{-1}). An exact position may be obtained by wave-dispersed measurements, which are, however, difficult because of large polymer absorption at harmonic frequency.

The EFISHG measurements on thin film also yield information about the conduction-band position in solid polydiacetylynes. The fact that one observes a permanent electric polarization of fundamental wavelength smaller

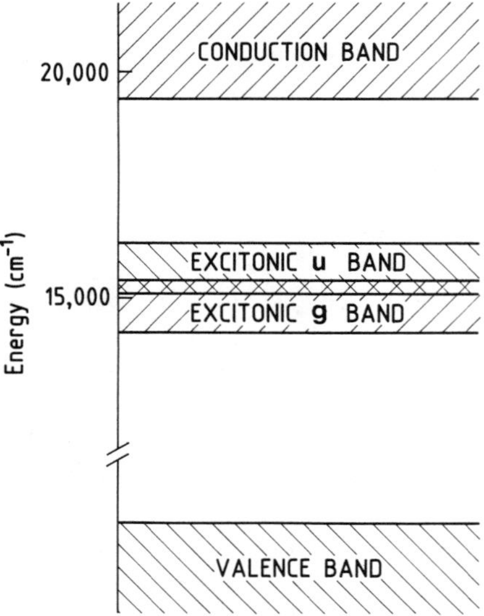

Fig. 14. Band structure of solid polydiacetylene blue form (g, two-photon; u, one-photon state).

than 1.1 μm, connected with electron–hole creation by two-photon absorption, allows us to place the conduction band at 18,500 cm^{-1} above the fundamental state and above the first optically allowed state (15,600 cm^{-1}). It means that the first optically allowed state is an excitonic band, in agreement with electroreflectance (Sebastian and Weiser, 1981) and photoconductivity (Lochner *et al.*, 1978) studies. Similarly, the two-photon state is also excitonic in origin. The resulting electronic structure of the blue form of LB films of polydiacetylene is displayed in Fig. 14.

B. Two-Photon Absorption

The existence of a two-photon state will be closely connected with the two-photon absorption (TPA) of the fundamental beam. This is due to the laser-induced refractive index variation at fundamental frequency ω,

$$n = n_0 + n_2 J^\omega \tag{69}$$

where

$$n_2 = 12\pi^2 \chi^{(3)}(-\omega; \omega, \omega, -\omega)/n_0 c \tag{70}$$

and n_0 is the linear refractive index, assumed to be real.

Although for the refractive index variation at frequency ω only the Kerr susceptibility $\chi^{(3)}(-\omega; \omega, \omega, -\omega)$ is responsible, this one will be also enhanced at the two-photon resonance:

$$\chi^{(3)}(-\omega; \omega, \omega, -\omega) \propto \frac{1}{E_{mg} - 2\omega - i\Gamma}\left[\frac{1}{(E_{ng} - \omega)^2}\right] \qquad (71)$$

In the vicinity of the two-photon resonance, this susceptibility will be complex, and consequently the refractive index will also be complex. Thus an intensity-dependent absorption of the fundamental beam will appear (TPA). This absorption, depending on $\mathrm{Im}\chi^{(3)}(-\omega; \omega, \omega, -\omega)$, will be the most important one at exactly the two-photon resonance position [$\mathrm{Re}\chi^{(3)} = 0$, cf. Eq. (67)] and will decrease going away from this resonance. We note here that this absorption is not connected in polydiacetylenes with electron–hole pair creation (around 1.35 μm).

C. Perspectives on Applications in Optical Devices

The excitonic origin of cubic susceptibilities in polydiacetylenes make them a promising candidate for applications in optical devices. It ensures a rapid response time (10^{-12}–10^{-14} sec). The fact that the excitonic band is separated from the conduction band allows us to avoid electron–hole pair creation, whose recombination time is generally long (10^{-6}–10^{-9} sec).

The discovery of the two-photon resonant state lying below the one-photon level in the blue form of LB films makes the wavelength of special interest for applications of these polymers. One can imagine use of thin LB films of polydiacetylene as frequency convertors in the near-infrared. The conversion efficiency for THG depends on the square of the incident laser-beam intensity.

After Lequime and Hermann (1977) and Ducuing (1977), the threshold damage for polydiacetylenes is larger than $50\ \mathrm{GW/cm^2}$ with picosecond pulses at 1.907 μm. In the vicinity of the two-photon resonance, the conversion efficiency will be limited by two-photon absorption, which probably will significantly decrease the damage threshold. In fact, as follows from Eqs. (67)–(68) and (71), the ratio

$$\frac{|\chi^{(3)}(-3\omega; \omega, \omega, \omega)|}{\mathrm{Im}\chi^{(3)}(-\omega; \omega, \omega, -\omega)} \approx [1 + (E_{mg} - 2\omega)^2/\Gamma^2]^{1/2} \qquad (72)$$

shows that it is impossible to get at the same time a large $\chi^{(3)}(-3\omega; \omega, \omega, \omega)$ susceptibility responsible for THG and a small two-photon absorption. The same remarks are also true for dc-induced SHG if one tends to use polydiacetylenes as optical or electro-optical switches. In order to avoid these difficulties, it is possible to choose the fundamental frequency outside of the two-photon resonance position, where the real part of $\chi^{(3)}$ reaches a maximum ($\omega = E_{mg} \pm \Gamma$). This is of special interest for the real part of the refractive index variation [depending on $\chi^{(3)}(-\omega; \omega, \omega, -\omega)$].

Two-photon resonance, which is harmful in frequency-conversion applications (for fundamental frequency close to the two-photon resonance energy E_{mg}), may be used in optical bistability, as discussed by Kothari and Kobayashi (1984).

The large permanent polarizations created in thin LB films may find eventual applications as dynamic electric memories. Writing of such memories can be done by the two-photon absorption process ($\lambda_\omega < 1.1$ μm), without damaging the thin film, and reading by use of a laser beam with a lower photon energy ($\lambda_\omega > 1.1$ μm), through SHG or other methods. Such memories can be stored in the dark a long time and erased again through the two- or one-photon process.

REFERENCES

Agrawal, G. P., Cojan, C., and Flytzanis, C. (1978). *Phys. Rev. B* **17,** 776.
Chance, R. R., Shand, M. L., Hogg, C., and Silbey, R. (1980). *Phys. Rev. B* **22,** 3540.
Chemla, D., and Kupecek, P. (1971). *Rev. Phys. Appl.* **6,** 31.
Chollet, P. A., Kajzar, F., and Messier, J. (1985). *NATO ASI Ser.* No. 102, 317.
Chollet, P. A., Kajzar, F., and Messier, J. (1986b). *Thin Solid Films* **132,**1.
Ducuing, J. (1977). *In* "Nonlinear Spectroscopy" (N. Bloembergen, ed.), pp. 276–295, North-Holland Publ., Amsterdam.
Genkin, V. M., and Mednis, P. (1968). *Sov. Phys.—JETP* (*Engl. Transl.*) **27,** 609.
Hanna, D. C., Yuratich, M. A., and Cotter, D. (1979). "Nonlinear Optics of Free Atoms and Molecules." Springer-Verlag, Berlin and New York.
Hudson, B., and Kohler, B. (1974). *Annu. Rev. Phys. Chem.* **25,** 437.
Jerphagnon, J., and Kurz, S. K. (1970). *J. Appl. Phys.* **41,** 1667.
Kajzar, F., and Messier, J. (1981). *Chem. Phys.* **63,** 123.
Kajzar, F., and Messier, J. (1983). *Thin Solid Films* **99,** 109.
Kajzar, F., and Messier, J. (1985a). (D. Bloor and R. R. Chance, eds.), *NATO ASI Ser.* No. 102, 325.
Kajzar, F., and Messier, J. (1985b). *Phys. Rev. A* **32,** 2352.
Kajzar, F., and Messier, J. (1986). *Thin Solid Films* **132,** 11.
Kajzar, F., Messier, J., Zyss, J., and Ledoux, I. (1983a). *Opt. Commun.* **45,** 133.
Kajzar, F., Messier, J., and Zyss, J. (1983b). *J. Phys.* (*Orsay, Fr.*) **44,** C3-709.
Kajzar, F., Ledoux, I., and Zyss, J. (1986). *Phys. Rev. A* (submitted for publication).
Kleinman, D. A., Ashkin, A., and Boyd, G. D. (1966). *Phys. Rev.* **145,** 338.
Kothari, N. C., and Kobayashi, T. (1984). *IEEE J. Quantum Electron.* **QE-20,** 418.
Langhof, P. W., Epstein, S. T., and Karplus, M. (1972). *Rev. Mod. Phys.* **44,** 602.
Lequime, M., and Hermann, J. (1977). *Chem. Phys.* **26,** 431.
Lieser, G., Tieke, B., and Wegner, G. (1980). *Thin Solid Films* **68,** 77.
Lochner, K., Baessler, M., Tieke, B., and Wagner, G. (1978). *Phys. Status Solidi B* **88,** 653.
Meredith, G. R., Buchalter, B., and Hanzlik, C. (1983a). *J. Chem. Phys.* **78,** 1533.

Meredith, G. R., Buchalter, B., and Hanzlik, C. (1983b). *J. Chem. Phys.* **78,** 1543.
Oudar, J. L. (1977). *J. Chem. Phys.* **67,** 446.
Rustagi, K. C., and Ducuing, J. (1974). *Opt. Commun.* **10,** 258.
Sauteret, C., Hermann, J. P., Frey, R., Pradere, F., Ducuing, J. Baughman, R. M., and Chance, R.R. (1976). *Phys. Rev. Lett.* **36,** 956.
Sebastian, L., and Weiser, G. (1981). *Phys. Rev. Lett* **46,** 1156.
Sewell, G. L. (1949). *Proc. Cambridge. Phillos. Soc.* **45,** 678.
Shand, M. L., and Chance, R. R. (1978). *J. Chem. Phys.* **69,** 4482.
Thalhammer, M., and Penzkofer, A. (1983). *Appl. Phys.* **B32,** 137.
Ward, J. F. (1965). *Rev. Mod. Phys.* **37,** 1.
Ward, J. F., and New, G.H.C. (1969). *Phys. Rev.* **185,** 57.

Chapter III-3

Degenerate Third-Order Nonlinear Optical Susceptibility of Polydiacetylenes

G. M. CARTER

MIT Lincoln Laboratory
Lexington, Massachusetts 02173

Y. J. CHEN, M. F. RUBNER, D. J. SANDMAN,
M. K. THAKUR, and S. K. TRIPATHY

GTE Laboratories, Inc.
Waltham, Massachusetts 02254

I. INTRODUCTION

A. Ultrafast Nonlinear Optical Signal Processing Based on $\chi^{(3)}$: Why Polydiacetylenes?

Several ultrafast nonlinear optical signal processing schemes rely on a material's intensity-dependent index of refraction as the basic nonlinear mechanism (Smith, 1982; Lattes *et al.*, 1983). In order to realize these concepts, a material is needed with both a large and a fast (subpicosecond response time) nonlinear optical coefficient, with the additional requirement that it be obtainable in the desired form (e.g., waveguides). An organic polymeric system, the polydiacetylenes (abbreviated as PDA) ($\overset{R'}{\underset{R}{>}}C{-}C{\equiv}C{-}C{<}$) (see

85

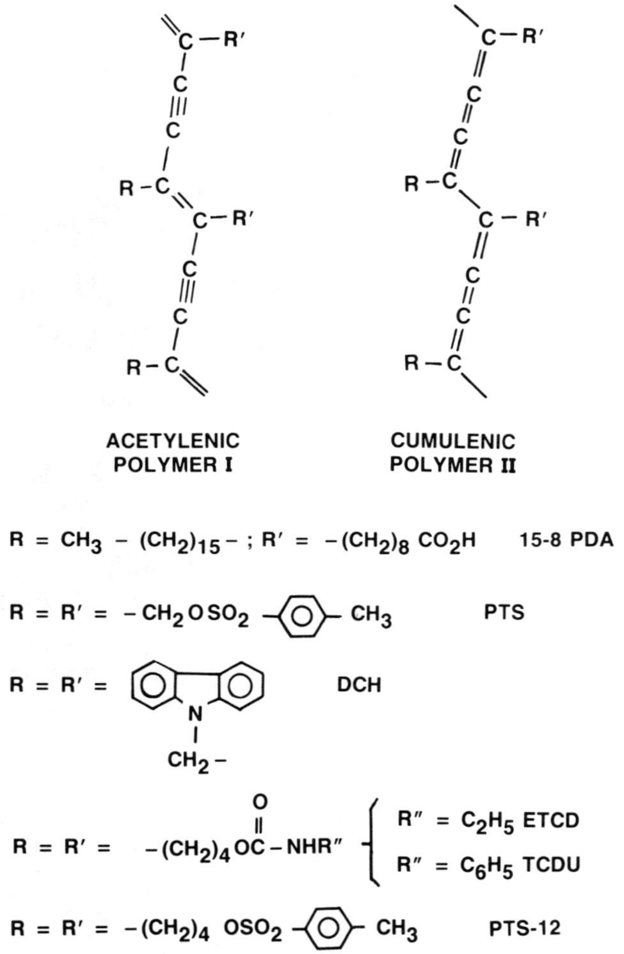

ACETYLENIC POLYMER I

CUMULENIC POLYMER II

R = CH$_3$ – (CH$_2$)$_{15}$ – ; R′ = –(CH$_2$)$_8$ CO$_2$H   15-8 PDA

R = R′ = –CH$_2$OSO$_2$ —⟨O⟩— CH$_3$ PTS

R = R′ = ⟨carbazole⟩ CH$_2$–   DCH

R = R′ = –(CH$_2$)$_4$OC(=O)–NHR″ { R″ = C$_2$H$_5$ ETCD ; R″ = C$_6$H$_5$ TCDU

R = R′ = –(CH$_2$)$_4$ OSO$_2$ —⟨O⟩— CH$_3$ PTS-12

Fig. 1. Bond representation and structural formulas for the polydiacetylenes discussed in this chapter.

Fig. 1), has attracted attention as a candidate material for such use (Smith, 1982) based on the following properties of the PDAs:

1.   The PDAs have one of the largest values of measured nonresonant third-order nonlinear optical susceptibility $\chi^{(3)}$, which is proportional to the intensity-dependent index of refraction (Hermann and Smith, 1980).

2.   The response time of the optical nonlinearity has been theoretically predicted to be subpicosecond (Smith, 1982).

3.   The PDAs are available in single-crystalline form for a large number of side groups R and R′ (which determine the polymer's organization).

4.   Because of this large variety of side groups, "molecular engineering"—i.e., a combination of proper choice of R and R′ through organic synthesis techniques—followed by material modification or processing, including crystal growth techniques, allows the desired material form to be obtained.

In this chapter we will review some of the basic physics of the optical nonlinearity, describe experimental results and materials research that demonstrate the nonlinear optical response of a polydiacetylene in planar waveguide form, and present related research in thin-film growth, materials characterization, and optical physics that is aimed at assessing the applicability of the polydiacetylenes for ultrafast signal processing.

B. Nonlinear Optical Properties of the Polydiacetylenes in the Solid State

The third-order nonlinear optical susceptibility $\chi^{(3)}(\omega)$ is the quantity that describes the change in dielectric constant ε_1 with optical intensity. In an isotropic material, ε, when expanded to the first nonlinear term, is (in esu)

$$\varepsilon = \varepsilon_1 + 4\pi\chi^{(3)}|E|^2 \tag{1}$$

where ε_1 is the linear dielectric constant of the material and E is the optical electric field. In general, the nonlinear contribution to the dielectric constant is much smaller than the linear one, allowing us to write for the index of refraction n of the material

$$n = n_1 + n_2 I \tag{2}$$

where n is the linear index of refraction and I is the optical intensity. Using Eq. (1), the coefficient n_2, which is defined as the nonlinear index of refraction in cgs units, is

$$n_2 = 16\chi^{(3)}/\pi^2 c\varepsilon_1 \tag{3}$$

where c is the speed of light.

The first measurements of $\chi^{(3)}(\omega)$ in a polydiacetylene was made in the PDA polybis (p-toulene sulfonate) of 2,4-hexadiyne-1,6 diol (PTS) at a wavelength of 1.9 μm by Hermann and Smith (1980). The value obtained is 3×10^{-9} esu and is the largest measured nonresonant $\chi^{(3)}(\omega)$. This value can be compared to another third-order process, third-harmonic generation. The third-order nonlinear susceptibility in general describes the generation of a fourth frequency, ω_4, due to three input frequencies, ω_1, ω_2, and ω_3, where $\omega_4 = \omega_1 \pm \omega_2 \pm \omega_3$, and is written as $\chi^{(3)}(\omega_4, \omega_1, \omega_2, \omega_3)$. When ω_1, ω_2, and ω_3 are all the same, ω, then $\chi^{(3)}(\omega)$ [Eq. (1)] is described by $\omega_4 = \omega + \omega - \omega$, i.e., equal to ω. Third-harmonic generation is described by $\chi^{(3)}(3\omega)$, where $\omega_4 = \omega + \omega + \omega$ is equal to 3ω. For electronic processes and for frequencies (ω and 3ω) below resonance, $\chi^{(3)}(\omega)$ and $\chi^{(3)}(3\omega)$ should have the same value. For PTS, using a laser wavelength of 1.9 μm, $\chi^{(3)}(3\omega) = 8 \times 10^{-10}$ esu (Sauteret *et al.*, 1976) (resonance wavelength = 6300 Å). Even though there is a factor of four discrepancy between these values, they are in reasonable agreement if one considers the large (60%) quoted errors for the $\chi^{(3)}(3\omega)$ measurements (Sauteret *et al.*, 1976). Theoretically, the large value for the nonresonant $\chi^{(3)}$ terms is attributed to the delocalized nature of the pi electrons along the carbon backbone of the one-dimensional polymeric system (Sauteret *et al.*, 1976; Agrawal *et al.*, 1978). On theoretical grounds (since the observed nonlinearity is due to the electronic states of the polymer, and since the wavelength used to measure $\chi^{(3)}(\omega)$ was below the absorption edge of PTS), the optical electric field used in Eq. (1) [and the optical intensity in Eq. (2)] is the instantaneous value. Thus the dielectric constant, and hence the index of refraction, can change as fast as the applied optical field—i.e., it adiabatically follows the field changes (Hanna *et al.*, 1979). This is the basis of using nonresonant optical nonlinearities for the ultrafast optical signal-processing schemes. Note that even though the one-dimensional PDA system is highly anisotropic, Eqs. (1)–(3) are still valid if all the optical fields are parallel to the carbon backbone direction.

C. Third-Order Nonlinear Optical Properties of the Polydiacetylenes in the Solid State: Directions for Research

Although the research described above has promising implications for nonlinear waveguide applications (Lattes *et al.*, 1983), there are several areas that need further investigation. There have been theoretical arguments that indicate that the strong coupling between the vibrational and electronic states for the carbon atoms of the PDA backbone may limit the response time of the optical nonlinearity to times longer than a psec (Flytzanis, 1983). However, the $\chi^{(3)}(\omega)$ measurement just mentioned (Hermann and Smith, 1980) has been carried out on the nanosecond time scale. Furthermore, for wavelengths near the absorption edge, the adiabatic arguments of the previous section break down (Hanna *et al.*, 1979), and the material parameters that determine whether or not one would observe the adiabatic effect at these wavelengths are unknown. The electronic states of the PDAs determine their nonlinear optical properties. Thus, one would like to have a description of these states in order to interpret the nonlinear optical experimental results as well as to provide a basis for material modification in order to enhance (enlarge) the value of $\chi^{(3)}$ in these systems. One model for the electronic states of the PDAs used to predict their nonlinear optical properties is the one-dimensional semiconductor model (Agrawal *et al.*, 1978), but this model is inconsistent with some photoconductivity measurements in these polymers (Pope and Swenberg, 1982).

In the following sections we will describe the interdisciplinary research effort in our laboratory that addresses some of these issues. The research discussed has a central goal: to help develop and assess the potential of the polydiacetylenes as a nonlinear optical material. The overall approach has been to simultaneously address "molecular engineering" and optical physics issues in these materials, because the structural and optical properties of the materials are interrelated. Research in one aspect of the work is very often driven by requirements, constraints, etc. in other areas. In Section II we describe research that led to the demonstration of nonlinear optical effects in planar waveguides of a polydiacetylene. This work is a good example of the synergistic output of multidisciplinary research. The Langmuir–Blodgett deposition technique was used to produce multilayers of a polydiacetylene in the technologically interesting form of a planar waveguide. By depositing a multilayer film on a holographic grating, coupling to the linear waveguide modes was observed. The value of n_2 for the multilayer film was measured using the same waveguide geometry. Various wavelengths were used to provide wavelength resolved nonlinear optical data near the material's absorption edge. In Section III we describe the research that was stimulated by these initial results. First, we describe improvements in the multilayer growth process and demonstrate via optical absorption and surface-enhanced Raman scattering data the ability to control the growth process. In parallel efforts, we present work aimed at providing a description of the ground and excited states in the polydiacetylenes. Evidence for the ground-state structure of these materials is presented. New crystal growth techniques that provide good optical quality material in thin platelet form are then described. These crystals provided us with the ability to probe the excited state in the crystals with four-wave mixing experiments. The resulting data provide the first information on the time response of excited-state optical processes in the polydiacetylenes (for wavelengths near the absorption edge) and indicate that the response time for these processes is less than 6 psec. A concluding section briefly summarizes the principal conclusions obtained from this research and sketches future directions for the research.

II. NONLINEAR OPTICAL MEASUREMENTS IN MULTILAYERS OF POLYDIACETYLENES

In this section, we describe the fabrication and subsequent nonlinear optical measurements of a polydiacetylene system in a technologically interesting and relevant form, namely, a planar waveguide. The fabrication technique employs the use of a Langmuir–Blodgett film balance to grow multilayers of polydiacetylenes on a suitable substrate. The monomers used in such a technique are designed so that they are specifically amenable to the Langmuir–Blodgett growth. Synthetic manipulation of the side-group architecture (R and R′) open up the possibility of extensive property modification and enhancement of the optical nonlinearities in these systems. In the prototype system, planar waveguide structures of quality sufficient to observe coupling to the appropriate waveguide mode were fabricated. The nonlinear optical measurements carried out in these waveguide structures provide the first wavelength-resolved n_2 results near the absorption edge of the material.

In Section II.A, the methods of sample preparation, including the growth of multilayer structures and the preparation of the silver-coated grating substrate, are outlined. In Section II.B, details of the nonlinear measurements from the polydiacetylene structures are presented.

A. Materials Preparation and Thin-Film Growth
Leading to Multilayered Waveguides of Polydiacetylenes

While some of the early measurements of the nonlinear optical properties of polydiacetylenes were carried out with large single crystals, there were significant problems associated with the poor optical quality of the crystals. In fact, to reduce the problem of uneven surface steps and resulting extensive scattering from the crystal surfaces, index-matching fluids were used (Hermann and Smith, 1980). In order to avoid these problems associated with single crystals then available, we chose to employ the Langmuir–Blodgett (LB) technique for the growth of planar waveguide structures. The use of the Langmuir–Blodgett technique for the growth of multilayer polydiacetylene structures provides several advantages. Large-area thin films with excellent adhesion to the substrate and optically flat surfaces could be obtained, and the LB process is amenable to the deposition of a variety of molecularly engineered diacetylenes and other functionalized monomers. The waveguide structures that were grown were appropriate tools for investigating the nonlinear optical properties of a whole class of polydiacetylenes molecularly engineered to have the appropriate functionalities and the desired densities of the polydiacetylene chains. Furthermore, the waveguide form is one of interest for at least one of the ultrafast signal processing schemes (Lattes *et al.*, 1983).

The basic requirement for the monomer to be amenable for the Langmuir–Blodgett growth is that it must have a hydrophilic head group and a hydrophobic tail. The monomers thus designed can be spread on the air–water interphase of a Langmuir–Blodgett film balance to form a monomer monolayer. The monomer monolayer can be organized as a continuous film by applying a constant pressure to the moving barrier.

The monomer used in our investigations, nonacosa-10,12-diynoic acid (also referred to as 15-8 monomer), as shown in Fig. 1, belongs to a class of diacetylene where appropriate separation between the diacetylene chains can be achieved by simply changing the length of the alkyl segments. The details of the monomer synthesis have been outlined elsewhere (Day, 1980). Extensive purification of the monomer is carried out by repeated crystallization of the monomer from the solution in petroleum ether. This monomer is one of a class of diacetylenes that have been extensively studied for their film-forming abilities and structural details. Limited studies of their electronic structure and optical properties have also been carried out. The monomers are ideally suited for LB growth, and the monomers become appropriately poised for solid-state polymerization subsequent to the application of the requisite barrier pressure. The monomers can be spread using a good solvent like chloroform.

The film balance employed in our investigation was a commercially available (Lauda) system, appropriately modified to permit uniform ultraviolet (UV) exposure to the entire film surface. Radiation from a UV lamp ($\lambda = 254$ nm) is used to polymerize the monomer monolayer film at the air–water interface. As the film is polymerized, there is some reduction in the total surface area of the film. The polymerized film can be transferred onto a suitable substrate by dipping the substrate through the film and subsequently pulling it off as a constant film pressure is maintained. Thus, in each of the dipping cycles a bilayer of the polymerized film is deposited. Successive deposition of several of these bilayers leads to the formation of the multilayer assembly. The preferred substrate for the formation of these multilayer assemblies is one with a strongly hydrophobic substrate, such as silver, which promotes strong adhesion between the surface and the hydrophobic tails of the molecules.

For the investigation of the linear and nonlinear optical properties of the multilayer polydiacetylenes, a waveguide coupling technique was employed. The substrate selected for the fabrication of the waveguide structure was a silver-overcoated grating that was etched on a silicon-wafer substrate. The grating was produced by a combination of holographic imaging and lithographic techniques. The grating space used was 5000 Å, and the depth was ~ 500 Å. A 1-μm-thick layer of silver was subsequently vacuum-deposited. The multilayers assume the profile of the grating, since the layers are deposited one at a time. In order to form the waveguides, the total thickness of the deposited polymer film was ~ 5000 Å.

The planar waveguide structure assembled on a silver grating provides a very convenient means of assessing its optical properties. In addition to the ability of coupling into the waveguide, the silver grating helps achieve strong adhesion due to its hydrophobic characteristics. It also effectively reduces in half the thickness of the polydiacetylene waveguide structure required to support a waveguide mode by essentially reflecting the mode back onto itself.

B. Linear and Nonlinear Optical Measurements in Multilayers of Polydiacetylene in Planar Waveguide Form

Using the planar waveguide form for the nonlinear optical measurements has several advantages: (1) the samples are already in a form of technological interest, namely, an optical waveguide; and (2) a nonlinear measurement technique is applicable for measurements of this material form and has the sensitivity to measure values of $\chi^{(3)}$ ($> 10^{-10}$ esu), even in films as thin as 4000 Å (Carter *et al.*, 1983a). In addition, this technique can be readily applied to both the temporal (response time) and spectral (wavelength dependence) studies of the polymeric materials.

In this section we will briefly review linear waveguide theory and describe the linear optical measurements in the PDA planar waveguides and the nonlinear optical experiments. The measured values of n_2 as a function of wavelength λ will also be presented.

1. Linear Waveguide Properties

For simplicity, we will only consider planar waveguide systems, although the discussion is quite general and can be applied to other waveguide systems (e.g., channel waveguides) as well. For an asymmetrical planar waveguide as shown in Fig. 2, the dispersion relation of the waveguide is determined by the wavelength, the dielectric properties of the two media, the guiding layer and the substrate, and the thickness of the waveguide. We will assume that the third medium is air, in which the dielectric constant is independent of the light intensity and wavelength. The dispersion of the transverse electric (TE) waveguide modes of the system can be calculated using the expression (Yariv, 1975)

$$\tan(ht) = (p + q)/[h(1 - pq/h^2)] \tag{4}$$

where β is the propagation constant of the waveguide mode; $q^2 = (\beta^2 - n_1 \mathbf{k}^2)$,

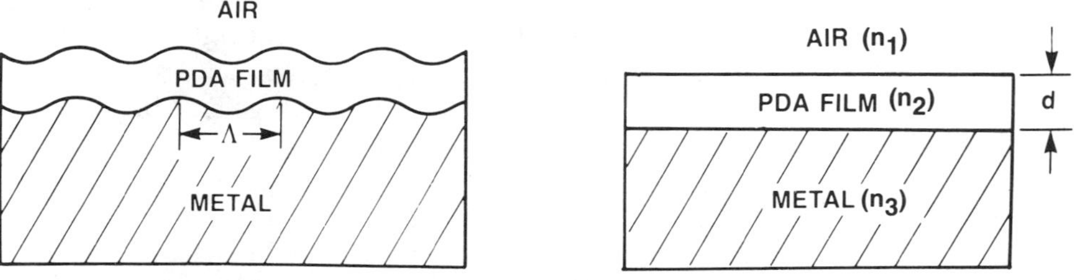

Fig. 2. Schematic representation of a planar air–LB polydiacetylene–silver asymmetrical waveguide.

$h^2 = (n_2^2 k^2 - \beta^2)$, and $p^2 = (\beta^2 - n_3^2 k^2)$ are the normal components of the wave vector of the media 1, 2, and 3, respectively; n_1, n_2, and n_3 are the indices of refraction of the media 1, 2, and 3, respectively; $\mathbf{k}$ the wave vector of the optical field in air; and t the thickness of the film. For transverse magnetic (TM) modes the dispersion relations are governed by the equation (Yariv, 1975)

$$\tan(ht) = [h(P + Q)]/[h^2 - PQ] \tag{5}$$

where $P = (n_2^2/n_3^2)p$ and $Q = (n_2^2/n_1^2)q$.

With this background, one can describe the waveguide properties of the PDA multilayer waveguides presented in the preceding section. As was described in Section II.A, the PDA multilayers were grown on a grating with silver overlayers. Silver provides a substrate of low (negative) index for the waveguide, and the grating allows one to couple optical energy into the waveguide and determine both the linear and the nonlinear intensity dependent waveguide properties. Figure 3 shows the schematic diagram of the apparatus used for these measurements. The laser source used was a Q-switched doubled Nd:YAG pumped dye laser system (100 mJ/pulse, 10 pulses/sec with a pulse width 5 nsec). The laser beam was incident at the polymer–air interface at an incident angle θ_i. The collimated and polarized

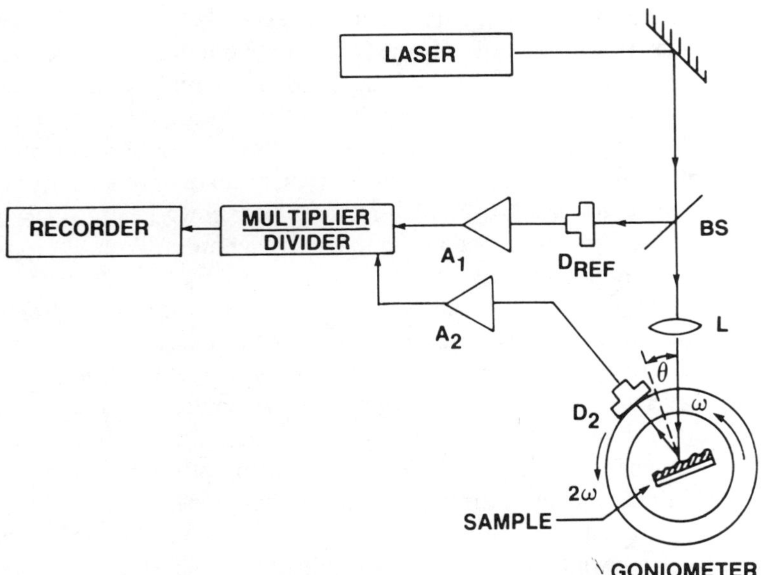

Fig. 3. Schematic diagram of the linear and nonlinear coupling setup. D_2 and D_{REF} are photodetectors, A_1 and A_2 are amplifiers, BS is the beam splitter, and L is the lens.

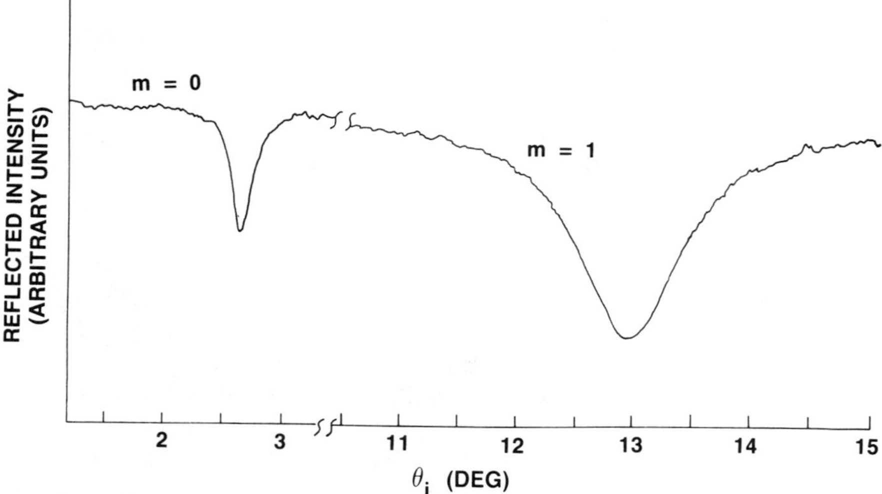

Fig. 4. Reflectivity versus incident angle for TE polarized light (7500 Å). Two TE waveguide modes were observed as designated.

laser beam is coupled to the waveguide mode(s) at the resonant angle θ_c when

$$\beta = nG \pm k\sin\theta_c \tag{6}$$

where n is the grating coupling order (an integer), $G = 1/\Lambda$, and Λ is the grating spacing. Thus the dispersion of the waveguide mode(s) can be readily determined by the measurement of the resonant coupling angle. In our experiments we detected the direct reflection of the incident laser beam from the waveguide structure, and thus in the experiments θ_c is the incident angle where the minimum in reflectivity occurs. A typical curve of reflectivity versus incident angle (θ_i) for s-polarized light at 7550 Å is shown in Fig. 4. Two TE waveguide modes were observed at the -1 coupling order. We want to point out that the experimental results shown in Fig. 4 not only provide the dispersion relation information of the waveguide modes of the air–polymer–silver system, but also provide information on the mode loss and grating coupling efficiency from the width and the size of the reflectivity minimum. Since the wavelength of the incident light is below the absorption edge of the 15-8 PDA, the mode loss is mainly due to the presence of the grating at the two interfaces (surface roughness) and the inhomogeneity of the polymer film. In fact, since the electric field distributions at the two interfaces (PDA–air and PDA–silver) are different for the two TE modes, from the difference in coupling efficiency and mode loss we determined that the PDA–air interface dominates the grating coupling effect.

Figure 5 shows the resonant coupling angles versus wavelength of two TE modes for a 5000-Å 15-8 PDA planar waveguide. From these curves we determined that the grating coupling order is -1. Since the polymer film thickness is fixed, one can also use the results shown in Fig. 5 to obtain the linear optical index of refraction of the PDA film. Since we found that the domain size of the LB PDA film is approximately a few micrometers square (which is much smaller than the incident beam), it can be assumed that the polycrystalline polymer film is isotropic in the plane of the film. One can then readily obtain a least-square fit of the index of refraction of the PDA film versus wavelength, as well as the thickness of the film. We found that the index in this wavelength region is 1.5 ± 0.05. Also, we should point out that one can also apply the technique to study single-crystalline polymer films by taking into account the optical anisotropy that typically exists in these materials.

2. Nonlinear Optical Measurements

Since the propagation constants β of either TE or TM modes are functions of the index of refraction of the medium 2 and medium 3, when the index of refraction of either or both media is changed, so are the propagation constants of the waveguide. Therefore, by measuring the change of the propagation constants of the waveguide modes with light intensity, one can determine the third-order nonlinearity of the system [see Eq. (2)]. Obviously the analysis is simpler if the waveguide medium (medium 2) is the only nonlinear medium. The technique of using the nonlinear dispersion relationship to study the nonlinear optical properties was first developed to study the nonlinear surface plasmons at silver–Si and silver–GaAs interfaces (Chen and Carter, 1982), and our discussion here describes the extension of this technique to waveguides for measurements of the nonlinear optical properties of the Langmuir–Blodgett PDA films.

To obtain the nonlinear optical data, we varied the incident laser power, which varies the optical intensity in the waveguide and hence varies the propagation constant β. From Eq. (6), one would expect to observe a change in the resonant coupling angle θ_c. From Eq. (6) and for small changes in coupling angle, one can show that $\cos \theta_c \, \Delta\theta_c = n_2 \, \Delta\langle I \rangle$, where $\Delta\langle I \rangle$ is the change in "average" intensity in the waveguide mode. The "average" is really a weighted average performed over the waveguide cross section (Carter and Chen, 1983). Thus, by measuring the incident laser energy, pulse width, and spot size on the sample, and by calculating the power coupled in the waveguide mode relative to the incident power, and by determining the linear waveguide properties, one can calculate $\Delta\langle I \rangle$ and, therefore, n_2 from the data (Carter et al., 1983b). Figure 6 shows the magnitude of n_2 versus the incident wavelength. Further,

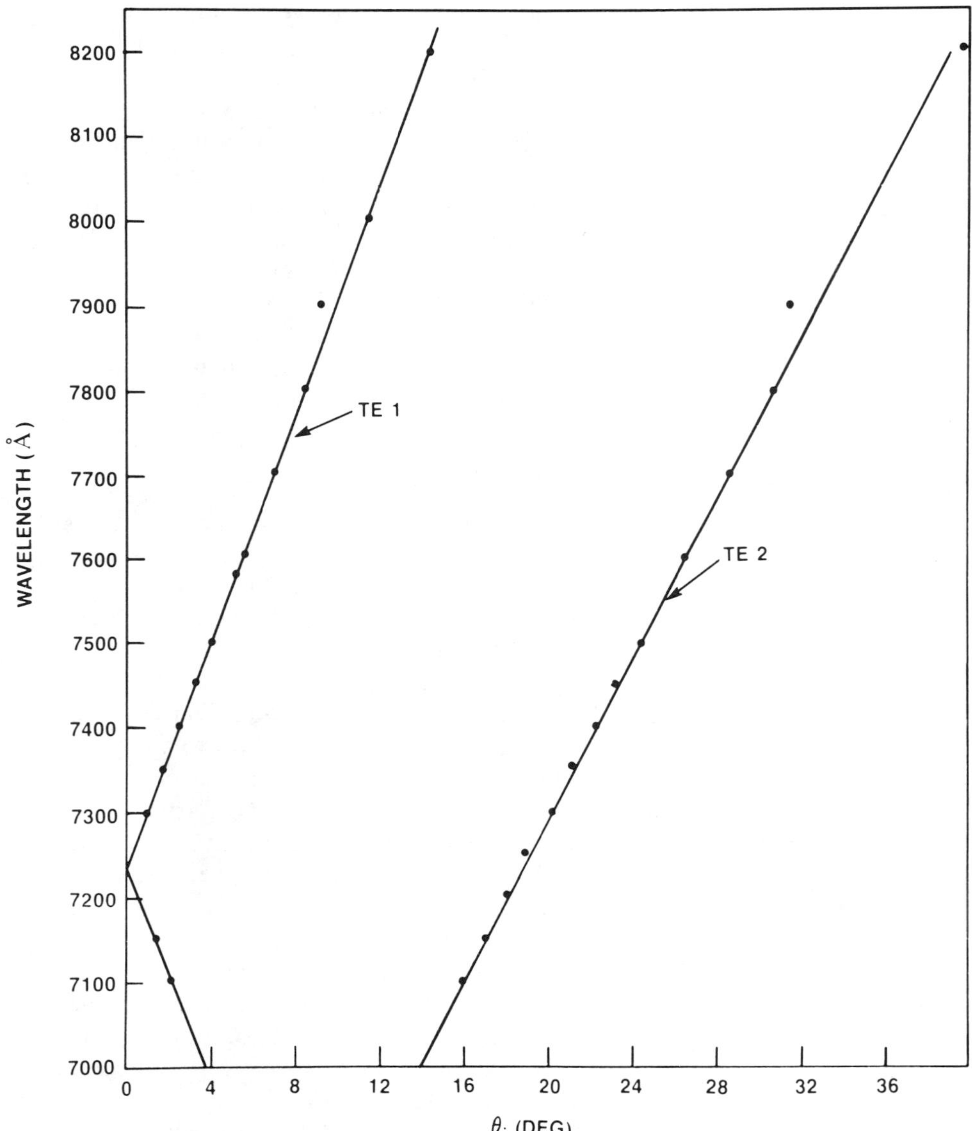

Fig. 5. Measured resonant coupling angle versus incident wavelength for the two observed TE waveguide modes in the ~5000-Å-thick multilayer polydiacetylene planar waveguide structure.

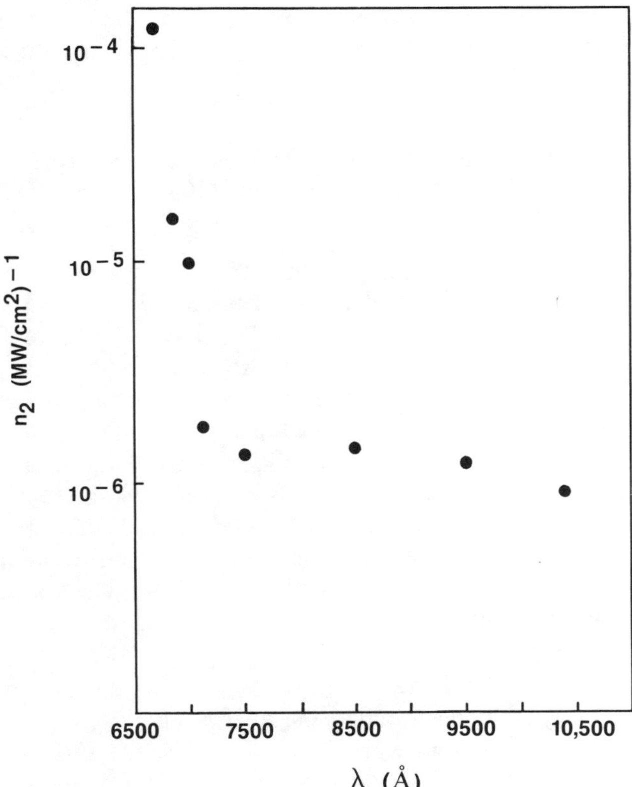

Fig. 6. Measured magnitude of n_2 for the multilayer polydiacetylene film in planar waveguide form as a a function of wavelength. The increase in the value for wavelengths shorter than 7500 Å is due to absorptive enhancement of the nonlinear process.

$\Delta n_g [n_g \equiv (\beta\lambda)/2\pi]$ was negative for all the data. Since we use the -1 grating coupling order, a negative Δn_g means n_2 is negative. The sign is opposite to that predicted by the one-dimensional semiconductor model for the polydiacetylenes (Agrawal *et al.*, 1978). The magnitude of the nonresonant $n_2(\lambda > 7000$ Å) in the PDA films prepared our laboratory, when corrected for the random ordering of the chains, is of the same order of magnitude (Hermann and Smith, 1980) and sign (P.W. Smith, private communication, 1984) as that measured by a different technique in single-crystal PTS at 1.9 μm. Thus, our data suggests that the nonresonant value of n_2 is relatively constant until one reaches the absorption edge of the film. This observation, coupled with the recent time-resolved n_2 measurements (see Section III.E) that have shown that these n_2 terms indeed have an ultrafast response time (<6 psec), implies that one should in principle be able to use these materials for the ultrafast signal-processing schemes even near their resonant wavelengths.

III. POLYDIACETYLENE MATERIALS RESEARCH AND TIME-RESOLVED NONLINEAR OPTICAL MEASUREMENTS

The research presented in the previous section concerned forming multi-layers of polydiacetylene into a planar waveguide and obtaining wavelength-resolved values of $\chi^{(3)}$ in these materials. The first two part of this section extend this work by examining the relationship between the material's structural and optical properties with emphasis on ways to optimize the linear and nonlinear optical properties of these films. The multilayer growth process can, in principle, be improved to provide a larger optical nonlinearity. First we discuss methods for obtaining (1) higher optical density and (2) single "blue" phase films, both of which on theoretical grounds should make $\chi^{(3)}$ larger than that observed in Section II. Next, surface-enhanced and resonance Raman data for these multilayers are analyzed. The results provide new insights into the Langmuir–Blodgett growth processes and help one understand the structural properties of these materials.

As discussed in Section I, knowledge of the ground-state and excited-state processes is a key to understanding the microscopic origins of $\chi^{(3)}$ in the polydiacetylenes. The next three articles concern this area. First, experimental evidence for the acetylenic ground-state representation in the polydiacetylenes is presented. Then, new single crystal growth techniques developed in our laboratory are reviewed. These thin platelets are ideal for probing the excited-state properties of the polydiacetylenes. In the final segment, measurement of the optical nonlinearity via time-resolved four-wave mixing in PTS polydi-acetylene platelets is presented. An ultrafast optical nonlinearity is found to be the dominant contribution to $\chi^{(3)}$ in this material, well into the absorption edge.

A. Manipulation and Polymerization of the 15-8 Diacetylene Monocarboxylic Acid Using the Langmuir–Blodgett Technique

In order to fully exploit the unique nonlinear optical behavior of the polydiacetylenes, it is important to establish the molecular and structural features of these materials that influence their linear and nonlinear optical properties. For the diacetylene monocarboxylic acids used to prepare monolayers and multilayers of controlled-thickness polydiacetylene thin films, it has been clearly demonstrated (Lieser *et al.*, 1980; Tieke *et al.*, 1979) by spectroscopic methods and electron and X-ray diffraction techniques that numerous phase changes can occur in both the monomer and polymer films during and subsequent to polymerization of these films to the optically active conjugated polymer. Accompanying these phase changes are dramatic shifts in the position of the absorption edge of the conjugated backbone, which is sensitive to the organization of the long-chain paraffinic side groups of the polymer. Such shifts are expected to have a profound effect on the nonlinear susceptibility of these materials, which is strongly dependent on the de-localization length of the pi electrons of the conjugated backbone. Thus, any changes in the electronic states of the backbone that are induced either reversibly or irreversibly during evaluation of the optical properties of the multilayers must be considered to properly assess the influence of structure and organization on the nonlinear optical behavior of these novel materials.

In the case of the 15-8 monomer, polymerization of a multilayer assembly or a solution-cast film of the monomer results in a polymer with an absorption spectrum with maxima at 640 and 580 nm (see Fig. 7a). This "blue" phase of the polydiacetylene is believed to grow within a monomer-rich matrix with only small changes in the packing of the molecules occurring during polymerization. Thus, for the most part, the order and orientation of the original organization is preserved as long as the polymer chains are maintained within the monomer modification. If, however, the multilayer is heated to 90°C or treated with suitable solvents, the "blue" phase is rapidly and irreversibly converted to the "red" phase, which exhibits absorption maxima at 540 and 500 nm (see Fig. 7b). In addition, the "red" phase exhibits a strong fluorescence, as shown in Fig. 8. Significant emission in amphiphilic polydi-acetylenes has only been reported for the "red-phase" polymer, which is believed to be highly disordered (Bubeck *et al.*, 1982). Thus, the structural changes that take place during the phase transition from "blue" to "red" introduce disorder into the polymer, in agreement with X-ray and electron-diffraction studies (Lieser *et al.*, 1980).

There are two possible general techniques that can be used to prepare LB multilayer assemblies of the surface-active polydiacetylenes. In the first technique, the monomer is organized at the air–water interface and sub-sequently deposited a monolayer at a time on a suitable substrate until the desired thickness is obtained. In this case, the final multilayer assembly is polymerized using UV radiation, resulting in a thin film with the optical characteristics of the "blue" phase of the polymer. This technique has been extensively utilized by many researchers (Lieser *et al.*, 1980; Tieke *et al.*, 1979; Bubeck *et al.*, 1982). The second technique involves polymerization of the organized monomer monolayer at the air–water interface, followed by transfer of the polymerized monolayer to a substrate using established

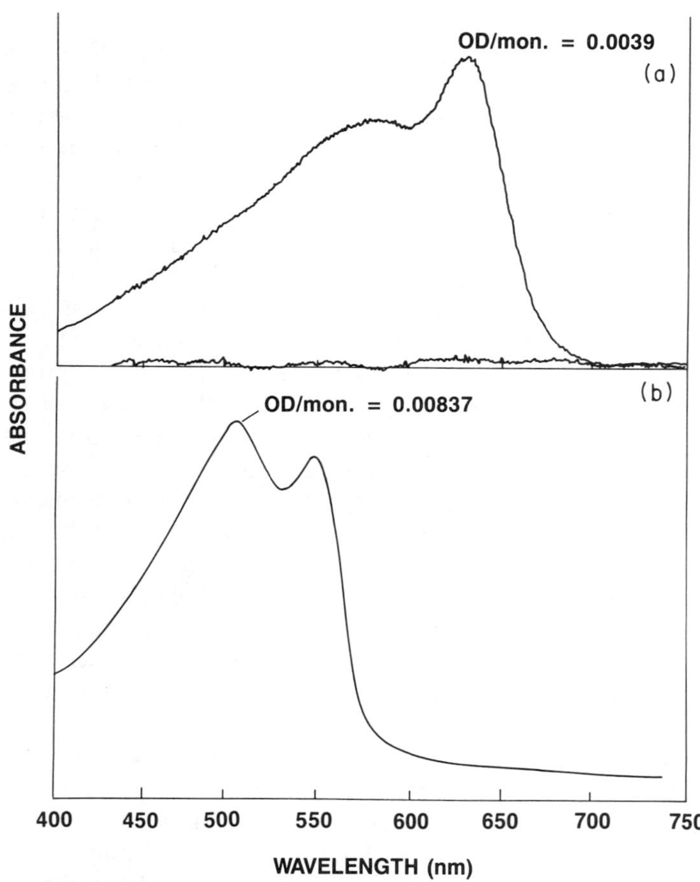

Fig. 7. Visible absorption spectra of (a) the "blue phase" for the 15-8 polymer and (b) the "red phase" of the 15-8 polymer.

deposition techniques. This process has been shown to produce an essentially "red" phase of the polymer unless special care is taken to limit the exposure time to the UV source (Day and Ringsdorf, 1979; Day *et al.*, 1979). For the 15-8 monomer it is difficult to obtain a "blue-phase" polymer by polymerizing at the air–water interface, since the polymerization proceeds with rapid conversion of the "blue" phase to the "red" phase as the extent of polymerization is increased. Thus, one is limited to lower conversions, and hence a smaller contribution of the conjugated backbone to the nonlinearity of the polymer if the "blue" phase is isolated and desired. This appears to be related to the greater degree of freedom experienced by the monomer at the air–water interface as compared to the more restrictive environment found in the monomer multilayer assembly (or solution-cast films). This is an excellent

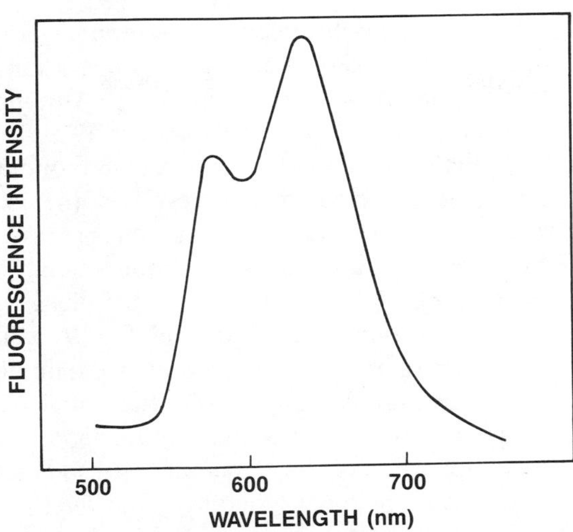

Fig. 8. Emission spectrum of the "red-phase" 15-8 polymer. The excitation wavelength for this data is 468 nm.

illustration of the unique behavior that can be found in two-dimensional monolayer films. The desire to produce optical-quality thin films for the evaluation of third-order effects in the polydiacetylenes has made it necessary to exploit the technique that introduces the least amount of defects and disorder in the film. We have found that polymerization at the air–water interface to a continuous polymer film followed by transfer to a substrate results in higher quality films as compared to the monomer transfer technique. However, to utilize this technique to produce the 15-8 polymer in the "blue" phase, it is necessary to modify the process to allow for the peculiarities of the reactivity of the monolayer. The technique used involves first polymerizing the monomer monolayer via UV radiation (254 nm) with a power density of 1 mW/cm^2 at 12.5°C for 45 sec at the air–water interface, followed by transfer of the partially polymerized monolayer to the substrate. After the multilayer of the desired thickness was formed, the film was further polymerized with UV light for 5 min at room temperature with the same exposure intensity. The result was an all "blue-phase" polymer, with an optical density per monolayer of 0.004 at 630 nm. Characterization of these films is now underway to determine their quality.

Preparation of an all "red-phase" 15-8 polymer can be readily accomplished by UV polymerization of the monomer monolayer at 20°C for 9 min ($\lambda = 254$ nm, power density $= 1 \text{ mW/cm}^2$) at the air–water interface, followed by the transfer of the polymer monolayer to the substrate. To insure a complete "red" phase, the final multilayer can be heat-annealed at 70°C for a few minutes, resulting in a polymer with an optical density per monolayer of 0.008 at 500 nm. An interesting effect was observed during the preparation of an all "red-phase" 15-8 multilayer that is worthy of discussion. Although it has been established by previous workers (Day and Ringsdorf, 1979; Day *et al.*, 1979) and confirmed by us that polymerization of the 15-8 monomer at the air–water interface for extended periods of time results in the completely "red" phase of the polydiacetylene, we have found that in the process of transferring the "red-phase" polymer monolayer to a substrate, a small but significant percentage of the "red" phase is converted to "blue-phase" polymer. This mechanically induced transition is illustrated in Fig. 9, which shows absorption spectra of the 15-8 polymer as a function of the number of bilayers deposited on a glass substrate. As can be seen, the initial bilayer is in the completely "red" phase, but as subsequent bilayers are deposited, the presence of a small shoulder in the region of the "blue-phase" absorption band begins to appear. Thus, the original polymer monolayer and the initially deposited bilayer are free to adopt the less-strained backbone conformation of the "red" phase. The process of sequentially building up the multilayer structure, however, imposes enough stress on some of the polymer molecules to convert

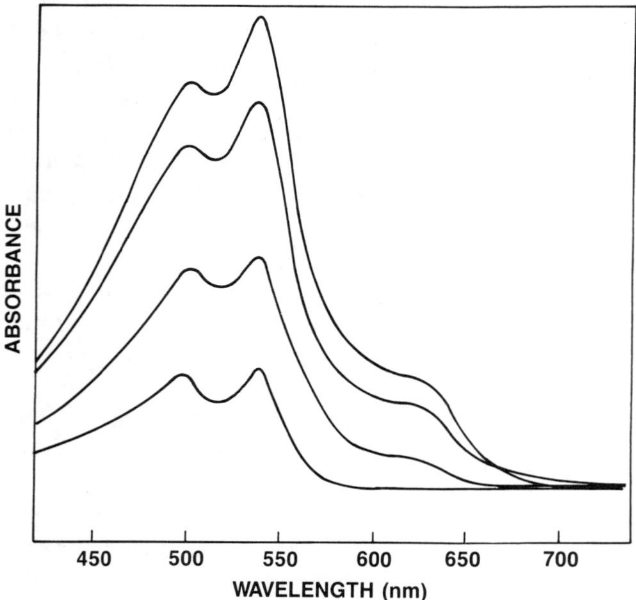

Fig. 9. Absorption spectra of the "red" phase for the 15-8 polymer for 1, 2, 3, and 4 bilayers transferred to a glass substrate. The lowest curve is for one bilayer, the next for two bilayers, etc.

them to the "blue" phase once the three-dimensional environment of the layered structure is established (>1 bilayer). This interpretation is supported by surface-enhanced and resonance Raman experiments described in the next section.

The ratio of the absorbance of the polymer at 630 nm ("blue-phase" region) and 500 nm ("red-phase" region) as a function of the number of bilayers transferred to the substrate is shown in Fig. 10. These data were obtained from three separate multilayer preparations and are quite reproducible. The conversion of a small percentage of the "red-phase" polymer to the "blue-phase" polymer demonstrates the unique role that the side groups play in influencing the electronic states of the backbone. We speculate that during the transfer of each monolayer, sufficient mechanical energy is imparted to selected regions of the multilayer film to place the conjugated backbone in its more planar conformation. The added strain imposed on the backbone most likely results from a reorganization of the side groups starting at the end groups present at the interlayer regions of the film. As can be seen in Fig. 10, absorbance in the "blue-phase" region increases relative to the "red-phase"

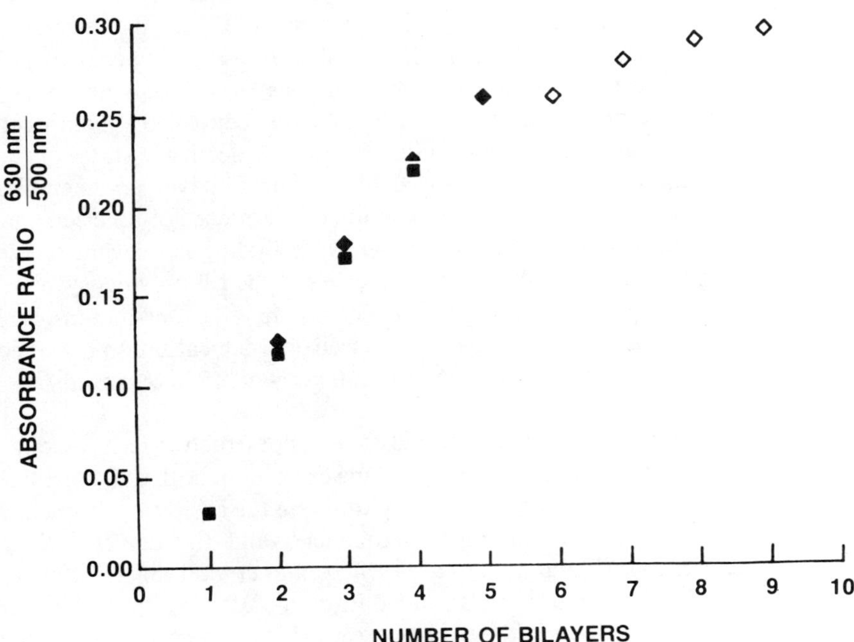

Fig. 10. The ratio of absorbance at 630 and 500 nm as a function of bilayers transferred to a glass substrate.

region until about 10 bilayers have been deposited, after which the ratio becomes nearly constant.

It is clear from the above discussion that the obtaining of thin films of optically pure phases of the surface active polydiacetylenes requires careful control over the number of different parameters of the LB film growth technique. Since each monomer has its own unique behavior in terms of organization and polymerization at the air–water interface, optimization of these parameters is required to obtain the desired optical properties.

B. Surface-Enhanced Raman Scattering

The preceding section examined the relationship between PDA's structural and optical properties and presented techniques for obtaining nearly single-phase multilayer films. In this section we report on the extention of this work by the use of surface-enhanced Raman scattering (SERS) to provide complimentary information concerning multilayer growth. By enhancing the Raman-scattering efficiency of the polymer films, we have observed for the first time the change of the structural and electronic properties of polydiacetylene LB films form monolayer (or one bilayer) to multilayers (Chen *et al.*, 1985). We found at the excitation wavelength of 6328 Å, near the resonance wavelength of the "blue-phase" PDA, a large shift ($50\,cm^{-1}$) to lower energy of the vibrational frequency of the stretching modes of the polydiacetylene backbone when two or more bilayers of LB films were deposited as compared to a monolayer or one bilayer. This indicates that the electronic states of the polymeric backbone become delocalized in the multi-bilayer polymer film system. The observation was the first experimental evidence of an electronic delocalization effect on the monolayer scale. This finding is relevant to the understanding of the nonlinear optical properties of the LB polymer systems, since the large optical nonlinearity observed in the one-dimensional organic polymers has been attributed to the electronic delocalization observed in the one-dimensional organic polymers (Sauteret *et al.*, 1976; Agrawal *et al.*, 1978).

Raman scattering is particularly useful as a nondestructive technique for studying large and small areas of LB films. One can obtain both the structural and the electronic properties of the film by studying the excitation frequency dependence of the vibrational mode frequencies and the corresponding Raman cross sections (Baughman *et al.*, 1974; Shand *et al.*, 1982; Batchelder and Bloor, 1984). The main limitation of Raman-scattering technique for studying thin polymer films is the detection sensitivity. One can increase the Raman-scattering efficiency by working near resonance and obtaining information on the states that were resonantly excited (Batchelder *et al.*, 1983).

Recently we have shown that by using SERS, one can probe the thin polymer films (down to one monolayer), even *off-resonance.*

There has been a considerable research activity in SERS since the observation of a "giant" enhancement ($\sim 10^6$) of the Raman scattering by molecules absorbed on a silver electrode (Jeanmaire and Van Duyne, 1977; Albrecht and Creighton, 1977). Although the phenomenon of SERS is still not completely understood, it is generally accepted that a major portion of the enhancement can be attributed to the field enhancement of the incident and scattered electromagnetic fields (Murray, 1983). The field enhancement effect, which can be explained as the local plasmon resonance at the rough metal surface, depends on the size and shape of the metal particles, the optical constant of the metal, and the position of the absorbed molecules on the metal surface. Although the electromagnetic fields generated by the dipolar plasma resonances of small metal particles are short-range, they have been used to enhance the Raman scattering by thin films on rough metal surfaces (Chen *et al.*, 1979; Murray and Allara, 1982).

The polydiacetylene monomer employed in our SERS experiment is the same material used in our nonlinear experiments described in Section II (15-8 PDA). The monomolecular layer of 15-8 PDA was completely polymerized on the LB trough by a long exposure (8 min) to UV radiation prior to the dipping process. The substrates were standard microscope glass slides. One side was vacuum-coated with a thick (~ 1 μm) smooth silver film, and the other side was vacuum-coated with a thin (mass thickness of $50-150$ Å) layer of silver film. Since the silver films are hydrophobic, two monolayers of polymer film were transferred to the silvered substrate during each dipping step.

Figure 11 shows the Raman spectra of one bilayer, two bilayers, and three bilayers of 15-8 PDA films on rough silver surface, with excitation at 6238 Å provided by a HeNe laser. The Raman spectrum of one monolayer film was very similar to that of a bilayer film, and the Raman spectra for thicker films, all termed multilayer films, are similar to that of two- or three-bilayers films. We note that the position of the two main Raman peaks, which are the $C{=}C$ and $C{\equiv}C$ stretch modes of the diacetylene backbone, are quite different for the one-bilayer film and the two- or three-bilayers films. As more bilayer films were deposited on top of the first bilayer films, the $C{=}C$ and $C{\equiv}C$ backbone stretch modes were shifted from 1521 cm^{-1} to 1456 cm^{-1}, and from 2123 cm^{-1} to 2078 cm^{-1}, respectively.

The shifts of the $C{=}C$ and $C{\equiv}C$ stretch mode frequencies toward lower energy can be explained by the electronic delocalization effect in the polydiacetylene backbone (Baughman *et al.*, 1974). The decrease of the bond orders of the $C{=}C$ and $C{\equiv}C$ bonds, due to the electron delocalization, can result in decreased vibrational frequencies for these bonds. The vibrational frequencies observed for the multilayer films correspond to that of the

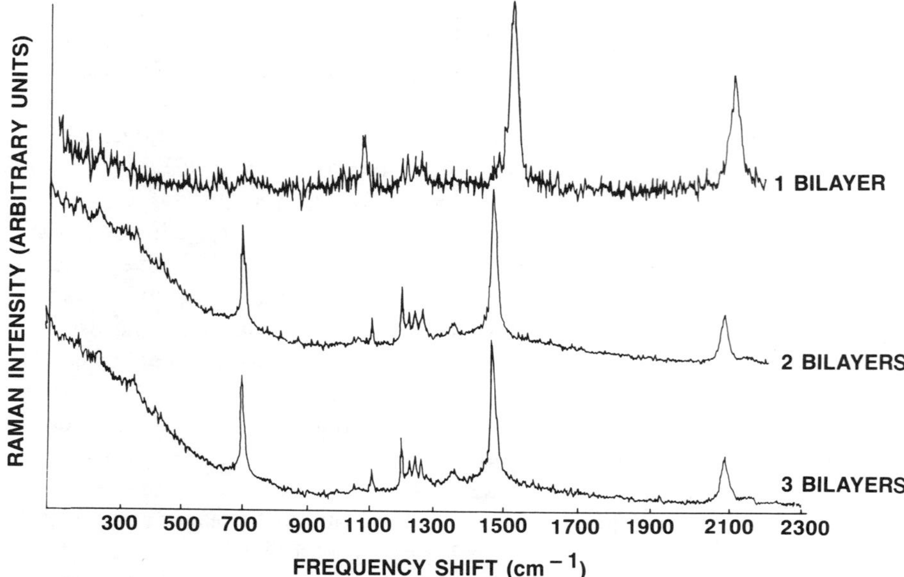

Fig. 11.   Raman spectra of one-, two-, and three-bilayer films measured by surface-enhanced Raman scattering with the incident wavelength 6328 Å.

"blue-phase" PDA (the ordered phase). Since the exciting wavelength, 6328 Å, is very close to the resonance wavelength of the "blue-phase" PDA, one would expect to observe these vibrational modes even if the multilayer film is a mixed-phase material. The two backbone stretch-mode frequencies observed for the one-bilayer (or monolayer) films correspond to that of a "red-phase" PDA (the less ordered phase). This suggests that the one-bilayer LB films deposited on the silvered surface are all "red-phase" materials.

We have also made a detail study of the LB films using the resonant Raman scattering technique. The results will be published in the future. We found that indeed the one-bilayer films are "red-phase materials," which have a resonant Raman frequency near 5400 Å. The multilayer films are mixed-phase materials that resonate at both the 6400-Å and the 5400-Å regions. When the multilayer films were excited near 5400 Å, the same stretch modes as the monolayer films were observed, indicating that the "red-phase" material is dominating the Raman spectra.

Although this work is continuing, we can already make some general observations based on the results just described. First, it appears that one bilayer is enough to act as a "buffer" layer, isolating subsequent layer growth from influences of the substrate. The subsequent growth is largely determined by the interlayer interactions. This interpretation is supported by the optical absorption data taken during multilayer growth presented in the previous section.

C. An Experimental Description of the Ground State of Polydiacetylenes

A complete understanding of the nonlinear optical properties of PDAs requires good descriptions of both the ground and the excited states of these conjugated polymers. For the ground state, desired information includes the absolute position of and data relevant to delocalized bonding in the occupied levels and details about the bond representations of the carbon atoms in the conjugated backbone. Since it is widely accepted that mechanical strains are created in the course of solid-state diacetylene polymerization (Baessler, 1984), experimental manifestations of such strains are clearly of interest. The experimental results to be summarized include results of studies of solid-state ionization energies (IP_c), single-crystal X-ray crystallography, ^{13}C nuclear magnetic resonance (NMR) spectroscopy, and vibrational spectroscopy.

Photoelectron spectroscopic techniques at ultraviolet (UPS) and X-ray (ESCA) energies have been widely used to study electronic structures of inorganic metals and semiconductors, as well as organic insulators and conductors (Grobman *et al.*, 1974; Nielsen *et al.*, 1974). These techniques were applied to PDAs by two research groups whose full papers appeared in 1978 (Stevens, *et al.*, 1978; Knecht and Baessler, 1978). From the data, one can estimate IP_c as 7 ± 1 eV and a valence bandwidth of 5 eV for poly-PTS. Subsequently, a photoionization technique using levitated microcrystals estimated for poly-PTS IP_c as 5.2 ± 0.1 eV (Arnold, 1982). Since such techniques use samples with a high density of surface defects, it was of interest to study photoemission with photons of energy 7–11 eV using single-crystal poly-PTS (Murashov *et al.*, 1982). A value of IP_c of 5.5 ± 0.1 eV was recorded for the polymer, significantly lower than that of PTS monomer, 7.1 ± 0.1 eV (Murashov *et al.*, 1982). Thus, significant delocalization occurs on polymerization. For poly-1, 6-di-*N*-carbazolyl 2,4-hexadiyne (poly-DCH), knowledge of the band gap and electron injection from magnesium to the bottom of the conduction band allowed estimation of IP_c as 5.8 ± 0.2 eV (Spannring and Baessler, 1981).

While single-crystal X-ray diffraction studies can give information about polymerization mechanism and polymer phase transitions, the major focus of these studies has been to deduce between structures **I** and **II** (see Fig. 1) as to which is the better bond representation for a given PDA. Recently, a comprehensive review of the crystallographic studies of diacetylene monomers and polymers has been given (Enkelmann, 1984). Within the limitations of the existing crystallographic data, structure **I**, the acetylenic structure, seems to best represent the PDA chain. Significant contributions from structure **II** have been proposed for three PDAs, including (poly bis-phenylurethane of 5,7-dodecadiyne-1,12-diol (TCDU). In these latter cases, the experimental evidence for **II** is not compelling due to problems in these crystals, such as incomplete polymerization or thermal decomposition. Studies of electron-density distribution have not been reported in PDA crystals to date.

From solution studies, it is readily apparent that ^{13}C-NMR spectra can easily distinguish acetylenic and cumulenic carbons, and a study of soluble PDAs with urethane side chains indicated the usual acetylenic structure (Babbitt and Patel, 1981). However, the properties of a material in solution are not necessarily relevant to those of the crystalline solid state. The first studies of PDAs by ^{13}C cross-polymerization (CP) magic-angle spinning (MAS) NMR techniques involved (poly bis(p-toulene sulfonate) of 5,7-dodecadiyne-1,12-diol (poly-PTS12) (Wenz *et al.*, 1984) and a partially crystalline sample (Havens *et al.*, 1984a). Subsequently, ^{13}C CP-MAS NMR methods were applied to poly-PTS and poly-TCDU (Havens *et al.*, 1984b). The poly-TCDU is of interest because of the uncertainties of the crystal structure analysis (Enkelmann, 1984; Enkelmann and Lando, 1978). X-Ray diffraction analysis of the samples used establishes that they are the known phases. Acetylenic resonances were observed at 106.4 and 104 ppm downfield of tetramethyl-silane (TMS) for poly-PTS and poly-TCDU, respectively (Sandman *et al.*, 1986). Both polymers lack resonances in the range 136–171 ppm assignable to an sp-hybridized butatrenic carbon, based on solution data (van Dongen *et al.*, 1973). The 1,1,4,4-tetraphenyl derivative is the first butatriene studied by ^{13}C CP-MAS techniques (Sandman *et al.*, 1986). For a sample established by X-ray diffraction to be the crystallographically defined phase (Berkovitch-Yellin and Leiserowitz, 1977; Berkovitch-Yellin *et al.*, 1974) of this compound, resonances at 124.1 and 152.9 ppm were assigned to the α and β cumulene carbons, respectively (van Dongen *et al.*, 1973).

The existing ^{13}C CP-MAS NMR data for PDAs indicate the usual acetylenic structure and provide no evidence for a significant contribution from a butatriene structure.

Vibrational spectra are a source of information concerning ground-state structure and bonding. In PDA, the vibrations of the backbone are not infrared-active, and infrared spectra of diacetylene monomers and polymers show only side-chain vibrations and are typically indistinguishable. A detailed review of resonance Raman spectroscopy of conjugated systems including PDAs appeared recently (Batchelder and Bloor, 1984). Four fundamental vibrations of the backbone, double- and triple-bond bending and stretching, are Raman-active, and ground-state force constants have been deduced. The frequencies of the four fundamentals are comparable to those of model molecular compounds. Applied external stress clearly affects both stretching vibrations (Batchelder and Bloor, 1984). The stretching vibrations shift to higher frequency as PTS monomer is converted to polymer. It is difficult to unequivocally decide whether these spectral changes are due to mechanical stress or the differing polymer chain lengths as polymerization proceeds (Batchelder and Bloor, 1984). Low-frequency $(0-300 \text{ cm}^{-1})$ Raman spectra of PTS monomer have been recorded as a function of conversion to polymer, and weak features at $35-45 \text{ cm}^{-1}$ have been tentatively assigned as phonon modes (Prasad *et al.*, 1982).

The change in the elastic constant in the polymer chain direction as a function of conversion from monomer to polymer has been reported for DCH (Enkelmann *et al.*, 1980).

The ground state of the PDA chain may be described as a delocalized one-dimensional band of significant width with a threshold solid-state ionization energy in the range $5-6$ eV. This experimental summary is compatible with current theoretical treatments (Karpfen, 1980; Bredas *et al.*, 1981, 1982). The solid-state ionization energy is comparable to easily ionized organic molecules such as tetrathiafulvalene (Nielsen *et al.*, 1974). Existing single-crystal X-ray structural studies and ^{13}C CP-MAS NMR work indicate that the acetylenic structure **I** is the presently preferred bond representation for the PDA backbone. At the present time, there is no credible evidence for the butatriene representation **II**, although it continues to have proponents in certain cases (Rickert *et al.*, 1983). While it is widely accepted that there is a mechanical strain on the PDA backbone (Baessler, 1984) brought about by a mismatch between monomer and polymeric lattices resulting in a volume contraction in the crystallographic unit cell upon polymerization (Enkelmann, 1984), it is not apparent what is a clear experimental manifestation of the strain.

In order to calculate the nonlinear coefficients of the polydiacetylenes as well as understand how to enhance them, one needs a description of the ground- and excited-state wave functions for the materials. Realizing that this goal requires further theoretical and experimental research, recognizing that one needs "good" optical quality materials to study their optical properties and that to probe the excited state of highly absorbing materials, like the PDAs, thin samples are needed, we initiated research in thin single-crystal growth. The next two sections describe new growth techniques and time-resolved four-wave mixing in these crystals.

D. Thin-Film Crystal Growth and Characterization of Some Polydiacetylenes

The fabrication of polydiacetylene crystals is an important area of research in view of their promise for a variety of opto-electronic and nonlinear optical applications. There are a few major criteria to be satisfied for any nonlinear optical material to be suitable for any practical application. These are: (1) the single crystals must be large and free of bulk defects, (2) the single crystals should have optically flat surfaces, (3) the dimensions of the crystals should be controllable, and (4) the material should be transparent over the wavelength range of interest determined by compatibility to optical communication bands (~ 0.7–$1.5\ \mu$m). Polydiacetylene crystals can be grown in reasonably large sizes, and they are transparent in the desired wavelength region. However, conventional crystal growth techniques, such as slow cooling or evaporation of solvent, do not provide any handle on the control of surface quality. The defect density of such crystals is very high, usually due to the fact that the fundamentals of crystallization and nucleation processes in these systems, and hence the appropriate controls, have not been developed. Additionally, limited control can be exercised on the dimensions of the crystals in the conventional growth techniques. Therefore, it is essential that some systematic method is established that would lead to defect-free polydiacetylene crystals that would be appropriate for more detailed and reliable experimental characterization as well as practical applications. As discussed in Section I, the most versatile form that would facilitate both experimental measurements as well as nonlinear optical applications is the form of a large-area ($> 1\ $mm^2) thin-film single crystal in the waveguide structure. Also, it is of interest to measure the dynamics of the excited state to determine such important physical parameters as the excited-state lifetime. Since the polydiacetylenes typically are highly absorbing (peak α) values $\approx 10^5\ $cm^{-1}), thin samples are required for such research.

Presently, epitaxy is a commonly used effective technique for the growth of inorganic thin-film single crystals. However, the method of epitaxy, or lattice-matching, is not effective in the case of large organic molecules simply because it is difficult to identify crystalline substrates that satisfy the corresponding lattice matching. Specifically, diacetylene monomer crystals that contain more than 60 atoms in each molecule have lattice spacings close to 20 Å or more along the molecular axes. It is difficult to identify crystalline materials, organic or inorganic, that have such large lattice spacings while having acceptable substrate quality. As a result, epitaxial methods have not been successful to date in producing large-area thin single-crystal films of polydiacetylenes (Rickert *et al.*, 1983). Clearly, thin-film crystal growth of polydiacetylenes is a relatively unexplored and difficult domain of investigation and does not lend itself to conventional methods.

Recently we have investigated some new approaches that seem to be well suited for rod-like molecules such as diacetylenes. The diacetylene monomer molecules, particularly those that are crystallizable, have the diacetylenic rods at the centers with some flexible side groups and slightly polar ends (see Fig. 1).

In addition, these molecules are fairly long (~ 25 Å) in the axial directions with lateral dimensions of approximately 5 Å in an extended conformation. In a mobile phase, such as melt, or saturated solution, such molecules are expected to have very small diffusion coefficients because of their size, shape, and intermolecular interactions. Thus, any ordered disturbance imposed on such molecules would tend to retain its influence for a reasonable length of time. Such retention times can be further prolonged if the mobile phase is placed in the interface of two opposing surfaces. These are the basic points that underlie the methods that we have recently exploited for the thin-film crystal growth of diacetylene monomers. In general, we call these methods shear growth techniques, since a shear is applied in order to introduce the preliminary orientation of the molecules. The basic experimental steps are quite simple. The selected monomer material is brought to a mobile phase by melting or dissolution in suitable solvents. The mobile phase is placed at the interface of two optically flat opposing surfaces. An appropriate magnitude of pressure is applied to this substrate–monomer–substrate assembly. Subsequently, one of the substrates is moved at a slow speed with respect to the other and thus provides a uniaxial shear to the diacetylene molecular assembly. Finally, the whole system is very slowly cooled, or the solvent is evaporated under a constant pressure until complete transformation to the solid state is achieved. The magnitude of the applied pressure dictates the final thickness of the single crystals.

We have applied the shear method to TCDU and bis-ethylurethane of 5,7-dodecadiyne-1,12-diol (ETCD) monomers in the melt phase and PTS monomers in solution. The thin film crystals of TCDU and ETCD were approximately 1 square centimeter in area with the thickness ~ 1 μm. Application of higher pressure ($\sim 2.4 \times 10^7$ dyn/cm^2) led to even thinner (~ 0.4 μm) single crystals. The PTS single crystals were ~ 15 square millimeters in area and 1 μm in thickness. Polymerization was effected using established means (UV, γ radiation, etc., heat) (Enklemann, 1984).

Characterization of the crystals was performed using optical microscopy, X-ray and electron diffraction, Fourier-transform infrared (FTIR), and optical absorption spectroscopy. The quality of these crystals was much superior to conventionally grown bulk crystals. The defect density was much less or negligible—specifically, the PTS single crystals that were grown by solution-shear technique showed no observable surface defects over an area of a few square millimeters. Optical micrographs of some of these crystals subsequent to polymerization are presented in Fig. 12. The large optical anisotropy of polydiacetylene single crystals is quite well known. Light polarized parallel to the chain axis is highly absorbed in the visible range, while for a perpendicular polarization the absorption is negligible. The optical absorption spectra of the thin-film poly-TCDU single crystals as a function of the orientation of

 G. M. Carter *et al.*

Fig. 12. Optical micrographs ($\times 200$) for thin-film single-crystal polydiacetylenes: (a) poly-TCDU (~ 1 cm^2 in area) obtained by melt-shear growth, and (b) PTS polydiacetylene (~ 50 mm^2 in area) grown by a solution-shear method.

Fig. 12.   (*Continued*)

polarization are shown in Fig. 13, which clearly demonstrates the large optical anisotropy typical of polydiacetylenes. The position of the absorption bands of the thin-film poly-TCDU and ETCD has been found to be quite different from the corresponding bulk crystals at room temperature. The first sharp absorption peak, commonly believed to be excitonic in origin, appeared at 605 nm (Fig. 13) for poly-TCDU thin film, instead of 550 nm as observed in the bulk crystals. In the case of poly-ETCD, the same peak appeared at 550 nm instead of 630 nm, as known for the room-temperature bulk crystals. In addition, the X-ray and electron diffraction studies have shown that the thin-film poly-TCDU and ETCD crystals have different unit-cell parameters compared to the corresponding bulk crystals. Thus, the melt-shear growth method has resulted in a new crystal phases for poly-TCDU and ETCD. However, the solution-shear growth in the case of PTS monomers did not give rise to any new crystal phase. Both the optical absorption properties and the unit cell parameters of the thin film PTS crystals are identical to that of the bulk crystals. Presumably, the more severe growth conditions in the melt-shear approach (pressure $\approx 1.6 \times 10^7$ dyn/cm^2 compared to the solution-shear method (pressure $\approx 4 \times 10^4$ dyn/cm^2) have led to the observed change in the crystal phases of poly-TCDU and ETCD. Details on the crystal structures of these new phases are presently under investigation. The X-ray diffractometer tracing of the thin-film single crystal of PTS polydiacetylene is

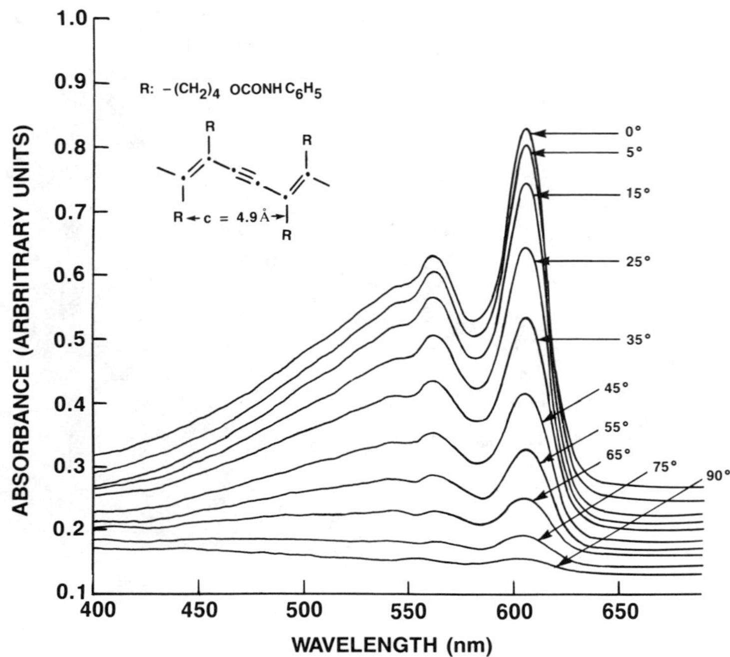

Fig. 13. Optical absorption spectra of a thin-film poly-TCDU crystal as a function of the angle of inclination of the incident polarization with respect to the polymer-chain axis. The large optical anisotropy is evident.

given in Fig. 14, which indicates that the side group (long axes) is nearly perpendicularly oriented on the surface of the substrates (Enklemann, 1984). Similar results on-side-group orientation have been obtained for poly-TCDU and ETCD as well.

The poly-ETCD bulk crystal is well known for its unusual nearly reversible thermochromism. The bulk crystals at room temperature are blue in color. Above 120°C they undergo a color change to red and subsequently reverse back to blue if the temperature is reduced. The thin-film single crystals, however, are pink in color at room temperature and do not show any thermochromism even at a temperature as high as 150°C, where they begin to decompose. Thus, the details of the chromic or optical properties of polydiacetylene crystals are closely related to the growth history. Since no difference in the chemical architecture of the molecules in the thin film and bulk poly-ETCD crystals was evident from the respective FTIR spectra, the difference in the chromic behavior was explained as being due to the difference in the packing of the side groups in the unit cells. The details of such structure–property relationships are still largely unknown and need more extensive investigation.

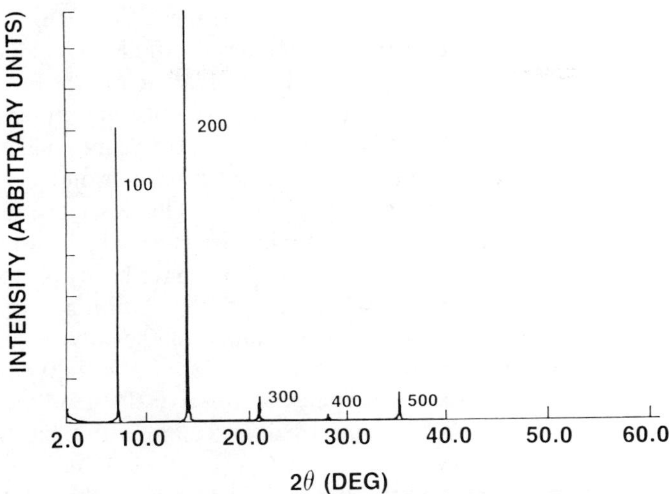

Fig. 14.   X-Ray diffractometer tracing of a thin-film single crystal of PTS polydiacetylene. The surface orientation was (100).

Clearly, with the development of these novel crystal growth techniques and the availability of thin-film defect-free single crystals, many new important and hitherto unattempted experiments now become possible for polydiacetylenes. However, along with the studies on the growth kinetics, careful characterizations of the resulting crystals in relation to the respective growth processes should be given equal importance.

E. Time-Resolved Degenerate Four-Wave Mixing Experiments in PTS Using Picosecond Laser Pulses

The single-crystal platelets of PTS described in the preceding section have good optical quality surfaces and produce even less surface scatter than the Langmuir–Blodgett–grown films described in Section II. Such crystals appear ideal for a variety of optical experiments designed to probe the excited state of these materials. In particular, we will describe a four-wave mixing experiment in the PTS platelets. The signal generated in these experiments is proportional to $|\chi^{(3)}|^2$ and hence provides another method for obtaining the third-order nonlinear optical coefficients in these materials. Further, the experiments were performed with short (picosecond) optical pulses and were designed to provide temporal information about the processes both in and near the material's absorption edge.

For the PTS platelets described (1 μm thick), one can estimate the amount of signal generated in a degenerate four-wave mixing experiment on the basis of the published values for $\chi^{(3)}$ in the material's transparent region. A simple calculation shows that for input power densities of 100 MW/cm², approximately 10^{-6} of the incident laser light would be converted into the fourth wave. This is on the order of the surface scatter for the PTS platelets at large angles from the incident beam. On the basis of this calculation we initiated a degenerate four-wave mixing experiment using pulses from a synchronously pumped cavity-dumped dye laser (Hryniewicz *et al.*, 1985). This laser produces high peak power ($\sim 1\,\text{kW}$) of short duration, typically 6 psec, which can be tuned from 5600 to 7900 Å depending on the dye used. We have been able to operate this laser system to produce pulses whose temporal duration nearly matches the Fourier transform of the spectrum over much of the wavelength range. Thus, by focusing the laser beam to a small spot size of about 20 μm, we can obtain enough power density to, in principle, observe the generated fourth wave. The experimental arrangement used is a standard phase-matched forward-degenerate four-wave mix geometry (Bogdan *et al.*, 1981) with two of the incident beams on translation stages to vary the relative time delay between the incident beams. Each of the beams is focused to the same 20-μm spot on the sample. At a incident wavelength in the absorption edge of the film, 6720 Å, we observed the generated fourth beam. Figure 15 shows the power in the fourth beam as a function of time delay where one of the beams is delayed relative to the other two (which are aligned in time). We only observed a signal when the three pulses overlapped in time, and verified

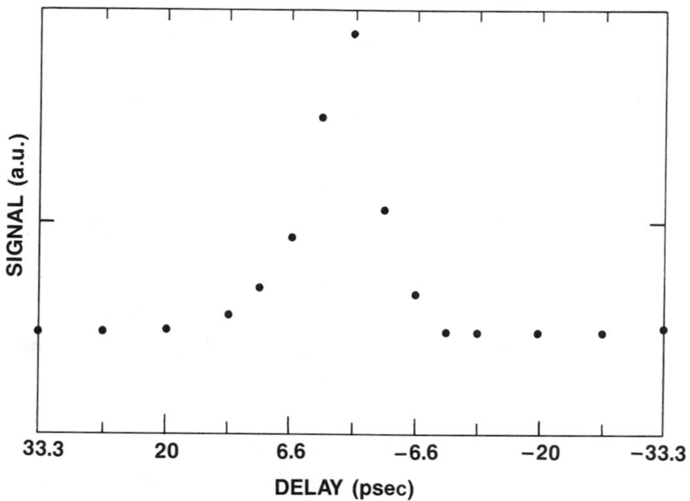

Fig. 15. Four-wave mixing signal in single crystal PTS polydiacetylene platelets. The delay is the relative delay between one of the incident beams and the other two. Positive delay means the "delayed" pulse arrives at the sample before the other two pulses. Laser wavelength is 6720 Å.

that the symmetric background was due to scatter. This, coupled with the fact that the four-wave mixing signal is narrower in time than the intensity autocorrelation of the laser pulses, indicates that the material's response time is less than our optical pulse width of 6 psec. Furthermore, by varying the input laser power, we have also verified that the power in the fourth wave is proportional to the cube of incident laser power, indicating a $\chi^{(3)}$ process since the three incident beams are derived from the same laser via beam-splitters. A preliminary measurement of the power generated in the fourth beam yields a value of $\chi^{(3)}$ in excess of three times larger than the long wavelength data taken on the nanosecond time scale. This indicates that there may be some resonant enhancement of $\chi^{(3)}$ in the absorption edge. Thus, even in the absorption edge, the nonlinear optical process that produces $\chi^{(3)}$ is ultrafast—i.e., its response time is shorter than our pulse width of 6 psec. This fast decay in the absorption region is the decay of the excited state, and recent experiments using 300-fsec duration laser pulses show the lifetime of the state to be 1.8 psec [G. M. Carter, private communication (1986)]. Furthermore, the magnitude of the process is consistent with a resonant enhancement of long-wavelength data taken on the nanosecond time scale. Experiments have been carried out to extend these measurements to longer wavelengths, and the nonresonant value of $\chi^{(3)} = 5 \times 10^{-10}$ esu at $\lambda = 7000\,\text{Å}$ (Carter *et al.*, 1985). Recent measurements using 300-fsec duration optical pulses have demonstrated that the response time of $\chi^{(3)}$ is shorter than the pulse duration [G. M. Carter, private communication (1986)].

Similar ultrafast temporal response for the four-wave mixing signal was observed in the Langmuir–Blodgett multilayers for wavelengths in their absorption region. However, the signal-to-noise ratio was drastically reduced because the multilayer samples are only $\frac{1}{2}$ μm thick and the chains are randomly oriented in the plane of the layers. However, this result is significant, and may imply that this ultrafast nonlinear optical process is a general property of all PDAs. Clearly, further four-wave mixing experiments on a variety of PDAs will establish the intimate relationship between the structural, electronic, and optical properties of these systems.

IV. CONCLUSION

As described in the introduction, the polydiacetylenes are candidate materials for ultrafast nonlinear optical schemes relying on the material's nonresonant third-order nonlinear optical susceptibility $\chi^{(3)}(\omega)$. The purpose of the interdisciplinary research presented in this chapter is to provide information and to develop techniques that will help realize these applications. In the area of optical physics we have made both wavelength- and time-resolved measurements of $\chi^{(3)}(\omega)$. These results show that the magnitude of $\chi^{(3)}$ is nearly constant from 1.05 μm to the material's absorption edge at ~ 7000 Å. At shorter wavelengths a large resonant enhancement is observed. From time-resolved picosecond four-wave mixing experiments in the material's absorption region, the response time of $\chi^{(3)}(\omega)$ was determined to be less than 6 psec. These results provide several important pieces of information. The fast response time coupled with the large nonresonant value of $\chi^{(3)}$ indicates that indeed the PDAs are attractive candidates for the nonlinear optical signal-processing application. This data also add to the progress in understanding of the microscopic origins of $\chi^{(3)}$—i.e., a description of the ground and excited states and their dynamics. The fast decay of $\chi^{(3)}$ in the absorptive region can be interpreted as a fast decay of the excited state initially populated by the laser. As discussed in the following, complementary experimental data is needed to verify this interpretation. In research concerning only the ground state, NMR measurements imply the ground-state structure for the PDAs to be acetylenic.

As pointed out in the introduction, materials research is also necessary so that one can obtain the form desired (e.g., a waveguide) for the proposed applications. By using the Langmuir–Blodgett technique, we demonstrated the existence of planer waveguide modes in a multilayer PDA structure. Further, the wavelength-resolved $\chi^{(3)}$ data were obtained in this technologically interesting structure. Surface-enhanced Raman scattering and optical spectroscopic techniques were used to study the multilayer formation process. The results established that the interaction with the substrate only occurred for the first bilayer deposited on the substrate. The phase or mixture of phases ("red" or "blue") of the multilayers is controlled by the bulk interactions. As part of this work, techniques for producing single "blue-phase" PDA multilayers were established. This is important because on theoretical grounds the "blue" phase is postulated to possess a much larger $\chi^{(3)}$ then the "red" phase. In a parallel effort, a novel solution-shear crystal growth technique was described. This effort yielded large-area platelets of PTS with good optical quality surfaces. Unlike the multilayers, the crystals have all of the chains aligned, which further increases $\chi^{(3)}$ and provides the ideal one-dimensional model systems. These thin crystals are ideal for studying the excited-state processes in the PDAs and were used in the picosecond four-wave mixing experiments.

The work presented in this chapter represents progress toward realizing optical applications using the polydiacetylenes. The information obtained also provides directions for future research that is necessary for further progress towards the applications. Ultimately, one would like to have the largest nonresonant value of $\chi^{(3)}$ possible. A more complete understanding of the microscopic origins of $\chi^{(3)}$ is needed so that one can manipulate ("molecular engineering") the material structure to enhance $\chi^{(3)}(\omega)$. In order to gain further information about the excited states, several experiments are necessary. For example, in a pump-probe type of experiment, an intense laser pulse can be used to populate an excited state, and the optical absorption at different wavelengths can be simultaneously probed. In this way one can monitor the time evolution of the population of the excited states. These experimental techniques have been demonstrated in other material systems with ~ 100 fsec time resolution (Fork *et al.*, 1983). Complementary information can be obtained by performing the four-wave mixing experiments with shorter (~ 100 fsec) optical pulses. In principle, one could observe the decay of the excited state populated by the laser by observing the asymmetry with respect to time delay in the four-wave mixing signal. In addition, further work in the material science area is indicated: (1) techniques are needed for obtaining multilayer films where all the chains are aligned; (2) methods for coupling energy into the single crystal platelets to demonstrate waveguiding in these samples should be pursued; and (3) in both the multilayer films and the single-crystal platelets, processing techniques are needed to define two-dimensional channel waveguides. Clearly the next decade of optical research in the polydiacetylenes and other molecular engineered organic materials will be as excited as the last!

REFERENCES

Agrawal, G. P., Cojan, C., and Flytzanis, C. (1978). *Phys. Rev. B* **17**, 776.
Albrecht, M. G., and Creighton, J. A. (1977). *J. Am. Chem. Soc.* **99**, 5215.
Arnold, S. (1982). *J. Chem. Phys.* **76**, 3842–3843.
Babbitt, G. E., and Patel, G. N. (1981). *Macromolecules* **14**, 554–557.
Baessler, H. (1984). *Adv. Polym. Sci.* **63**, 1–48.
Batchelder, D. N., and Bloor, D. (1984). *Adv. Infrared Raman Spectrosc.* **11**, 133–209.
Batchelder, D. N., Bloor, D., and Lyall, I. R. J. (1983). *Solid State Films* **99**, 118.
Baughman, R. H., Witt, J. D., and Yee, K. C. (1974). *J. Chem. Phys.* **60**, 4755.
Berkovitch-Yellin, Z., and Leiserowitz, L. (1977). *Acta Crystallogr., Sect. B* **B3**, 3657–3669.
Berkovitch-Yellin, Z., Lehav, M., and Leiserowitz, L. (1974). *J. Am. Chem. Soc.* **96**, 918–920.
Bogdan, A. R., Prior, Y., and Bloembergen, N. (1981). *Opt. Lett.* **6**, 82.
Bredas, J. L., Chance, R. R., Silbey, R., Nicholas, G., and Durant, P. (1981). *J. Chem. Phys.* **75**, 255.
Bredas, J. L., Chance, R. R., Baughman, R. H., and Silbey, R. (1982). *J. Chem. Phys.* **76**, 3673–3678.
Bubeck, C., Tieke, B., and Wegner, G. (1982). *Ber. Bunsenges. Phys. Chem.* **86**, 495.
Cantow, H. J. (1983). *Adv. Polym. Sci.* **63**.
Carter, G. M., and Chen, Y. J. (1983). *Appl. Phys. Lett.* **42**, 643.
Carter, G. M., Chen, Y. J., and Tripathy, S. K. (1983a). *In* "Nonlinear Optical Properties of Organic and Polymeric Materials" (D. J. Williams, ed.), Chapter 10. Am. Chem. Soc., Washington, D. C.

Carter, G. M., Chen, Y. J., and Tripathy, S. K. (1983b). *Appl. Phys. Lett.* **43**, 891.

Carter, G. M., Thakur, M. K., Chen, Y. J., and Hryniewicz, J. V. (1985). *Appl. Phys. Lett.* **47**, 457.

Chen, C. Y., Burstein, E., and Lundquist, S. (1979). *Solid State Commun.* **32**, 63.

Chen, Y. J., and Carter, G. M. (1982). *Appl. Phys. Lett.* **41**, 307.

Chen, Y. J., Carter, G. M., and Tripathy, S. K. (1985). *Solid State Commun.* **54**, 19.

Day, D. (1980). Ph.D. Thesis, Case Western Reserve University, Cleveland, Ohio.

Day, D., and Ringsdorf, H. (1979). *Makromol. Chem.* **180**, 1059.

Day, D., Hub, H., and Ringsdorf, H. (1979). *Isr. J. Chem.* **18**, 325.

Enkelmann, V. (1984). *Adv. Polym. Sci.* **63**, 91–136.

Enkelmann, V., and Lando, J. B. (1978). *Acta Crystallogr., Sect. B* **B34**, 2352–2353.

Enkelmann, V., Leyrer, R. J., Schleier, G., and Wegner, G. (1980). *J. Mater. Sci.* **15**, 168–176.

Flytzanis, C. (1983). *In* "Nonlinear Optical Properties of Organic and Polymeric Materials" (D. J. Williams, ed.), Chapter 8. Am. Chem. Soc. Washington, D. C.

Fork, R. L., Shank, C. V., Yen, R., and Hirlimann, C. A. (1983). *IEEE J. Quantum Electron.* **QE-19**, 500.

Grobman, W. D., Pollak, R. A., Eastman, D. E., Maas, E. T., Jr., and Scott, B. A. (1974). *Phys. Rev. Lett.* **32**, 534–537.

Hanna, D. C., Yuratich, M. A., and Cotter, D. (1979). "Nonlinear Optics of Free Atoms and Molecules," Chapter 2. Springer-Verlag, Berlin and New York.

Havens, J. R., Thakur, M., Lando, J. B., and Koenig, J. L. (1984a). *Macromolecules* **17**, 1071–1074.

Hermann, J. P., and Smith, P. W. (1980). *Dig. Tech. Pap.—Int. Quantum Electron. Conf., 11th, 1980,* pp. 656–657.

Hryniewicz, J. V., Carter, G. M., and Chen, Y. J. (1985). *Opt. Commun.* **54**, 230.

Jeanmaire, D. L., and Van Duyne, R. P. (1977). *Electroanal. Chem.* **84**, 1.

Karpfen, A. (1980). *J. Phys. C* **13**, 5673–5689.

Knecht, J., and Baessler, H. (1978). *Chem. Phys.* **33**, 179–183.

Lattes, A., Haus, H. A., Leonberger, F. J., and Ippen, E. P. (1983). *IEEE J. Quantum Electron.* **QE-19**, 1718.

Lieser, G., Tieke, B., and Wegner, G. (1980). *Thin Solid Films* **68**, 77.

Murashov, A. A., Silinsh, E. A., and Baessler, H. (1982). *Chem. Phys. Lett.* **93**, 148–150.

Murray, C. A. (1983). *In* "Recent Advances in Laser Spectroscopy" (B. A. Garetz and J. Lombardi, eds.), Vol. 4. Wiley, New York.

Murray, C. A., and Allara, D. L. (1982). *J. Chem. Phys.* **76**, 1290.

Nielsen, P., Epstein, A. J., and Sandman, D. J. (1974). *Solid State Commun.* **15**, 53–58.

Pope, M., and Swenberg, C. E. (1982). "Electronic Processes in Organic Crystals," pp. 677–696. Oxford Univ. Press, London and New York.

Prasad, P. N., Swiatkiewicz, J., and Eisenhardt, G. (1982). *Appl. Spectrosc. Rev.* **18**, 59–103.

Rickert, S. E., Lando, J. B., and Ching, S. (1983). *Mol. Cryst. Liq. Cryst.* **93**, 307–314.

Sandman, D. J., Tripathy, S. K., Elman, B. S., and Samuelson, L. A. (1986). *Synth. Met.* (to be published).

Sauteret, C., Hermann, J. P., Frey, R., Pradere, F., Ducuing, J., Baughman, R. H., and Chance, R. R. (1976). *Phys. Rev. Lett.* **36**, 956.

Shand, M. L., Chance, R. R., LePostollec, M., and Schott, M. (1982). *Phys. Rev. B* **25**, 4431.

Smith, P. W. (1982). *Bell Syst. Tech. J.* **61**, 1975.

Spannring, W., and Baessler, H. (1981). *Chem. Phys. Lett.* **84**, 54–58.

Stevens, G. C., Bloor, D., and Williams, P. M. (1978). *Chem. Phys.* **28**, 399–406.

Tieke, B., Lieser, G., and Wegner, G. (1979). *J. Polym. Sci., Polym. Chem. Ed.* **17**, 1631.

van Dongen, J. P. C. M., deBie, M. J. A., and Steur, R. (1973). *Tetrahedron Lett.* pp. 1371–1374.

Wenz, G., Muller, M. A., Schmidt, M., and Wegner, G. (1984). *Macromolecules* **17**, 837–850.

Yariv, A. (1975). "Quantum Electronics," 2nd ed., pp. 512–514. Wiley, New York.

Chapter III-4

Dimensionality Effects and Scaling Laws in Nonlinear Optical Susceptibilities

CHRISTOS FLYTZANIS

Laboratoire d' Optique Quantique, Ecole Polytechnique
91128-Palaiseau, Cedex, France
and Max Planck Institut für Quantenoptik
D-8046-Garching, Federal Republic of Germany

I. INTRODUCTION

In condensed matter the dependence of physical quantities, like susceptibilities, transport coefficients, and others, on the different microscopic parameters that describe the material system and its coupling with the external fields is quite complicate. Even in the rare cases where this is feasible, such detailed knowledge, in many respects, is cumbersome and even superfluous at first sight. This is in particular the case when one wishes to grasp general trends and derive interrelations between the different macroscopic quantities. In such cases a simple functional dependence of the macroscopic quantity on an effective parameter is far more useful and in many respects fundamental. The choice of the effective parameter is of central interest, as it must be both intuitively pertinent to the physical phenomenon considered and amenable to a rigorous definition within a quantum-mechanical description of the material

121

system; more importantly, it must also reflect the range of validity of a certain approach or model. The latter is indeed important at both a practical and a fundamental level, as quite often size effects become crucial in present-day technology to an extent that drastic changes occur both in the energy spectrum and wave functions and certainly in the quantum-mechanical averaged quantities that enter the expressions of physical quantities.

The usefulness of simple functional dependence of a physical quantity on an effective parameter relies on the fact that one can derive scaling laws of quite universal validity. Quite often these scaling laws are power laws with exponents that depend critically on the dimensionality of the material system for infinitely extended systems. If the extension of the material system is reduced to finite size beyond certain limits, the meaning of the effective parameter changes and the scaling laws change too or even break down. Reversing the line of thought, breakdown of a scaling law also indicates a change of the dimensionality or extension of the material system.

Dimensionality considerations and scaling laws have been extensively introduced in the theory of physical quantities like linear magnetic electrical or optical susceptibilities, but much less so for nonlinear optical coefficients. In a certain sense the first such direction was taken with Miller's empirical rule, which in its initial form states (Miller 1964) that

$$\chi^{(2)}_{ijk} = \delta_{ijk}\chi^{(1)}_{ii}\chi^{(1)}_{jj}\chi^{(1)}_{kk} \tag{1}$$

where $\chi^{(2)}$ is the second-order susceptibility and $\chi^{(1)}$ is the linear susceptibility; δ_{ijk} was conjectured to be approximately a constant by simple survey of the existant experimental values of $\chi^{(2)}$ for some 20 different crystals. This rule relates $\chi^{(2)}$ to $\chi^{(1)}$ where a large body of information exists. The first analysis of Miller's empirical rule using dimensionality arguments (Flytzanis and Ducuing, 1969) based on detailed quantum mechanical calculations of $\chi^{(2)}$ in covalent semiconductors revealed that quite generally δ is not a constant but roughly proportional to the electronic dipole moment of the electronic density distribution within a bond and even can reverse sign for a given axis convention.

Still, such an approach does not enter the spirit of scaling laws, as it relies heavily on a microscopic description. A fundamentally different approach, and more appropriate for extended periodic systems (crystals), to derive the functional dependence of $\chi^{(3)}$ on certain effective parameters without having recourse to any microscopic description in particular was used by Cardona and Pollack (1971), van Vechten et al. (1970), Agrawal and Flytzanis (1976), and Agrawal et al. (1978). There, use is made of the dimensionality character of the different critical points in the joint density of states in extended periodic systems (crystals) or van Hove singularities (see, for instance, Ashcroft and Mermin, 1981). Using a deductive approach, scaling laws were subsequently obtained for all odd-order susceptibilities (Flytzanis, 1983) in one-dimensional systems and related by simple power laws to the electron delocalization length.

We next present a general discussion of the scaling laws that seem to govern the nonlinear optical susceptibilities and polarizabilities. The effective parameter of central interest here is the electron delocalization length (Cojan *et al.*, 1977), which (as will be shown) can be simply expressed in terms of macroscopic quantities that can be measured.

At this level it is important to reduce the scope of the chapter and state the limits within which the laws are valid. Only the behavior of the nonlinear optical susceptibilities in the transparency region, the one extending below the lowest electronic transition and above the vibrational frequencies, will be considered. In particular, this amounts to restricting ourselves to the electronic contribution in $\chi^{(n)}$, the most relevant one for the problems we have in mind. By excluding resonances, we also expect that only the overall characteristics of the electron density distribution will be needed and not the detailed energy spectrum or the excited-state wave functions. We shall also assume that the energy separation between the highest occupied and lowest unoccupied electronic state is much higher than kT, so that no thermodynamic considerations will enter the picture.

With the exception of a short section, we will restrict the discussion to the odd-order dipolar susceptibilities $\chi^{(2n+1)}$. Under the physical conditions specified above, it will be shown below that the rather complex behavior of $\chi^{(2n+1)}$ can be reduced to simple power laws involving a single effective parameter, the electron delocalization length. This is the space available to an electron in a system with infinite extension and in general extends over several unit cells in a crystal; it is also the minimum "box" within which the electron can be squeezed without its states and response to external fields being affected by size effects. Indeed, as will be seen, this parameter can be simply expressed in terms of experimentally accessible quantities that are characteristic of the k-space (global quantities) rather than the r-space (local). The even-order dipolar susceptibilities $\chi^{(2n)}$ do not display such a simple structure. These are indeed sensitive to the charge asymmetry that is localized within a molecular entity (bond or molecule) and may interfere with the electron delocalization. Miller's empirical rule [Eq. (1)] gives a hint that the two effects can be factorized, but more work is still needed to obtain useful scaling rules of universal validity.

II. NONLINEAR OPTICAL SUSCEPTIBILITIES

In the subsequent sections we shall be interested in the behavior of the nonlinear optical susceptibilities for both small-size and infinite-extension systems, and we will set up here their general expressions.

For small-size systems (zero dimensionality), much smaller than the optical wavelength, the relevant quantity is the induced dipole p, whose dependence on the applied local field is expressed in terms of the dipolar polarizabilities $\alpha^{(1)}$, $\alpha^{(2)}$, etc. Using standard time-dependent perturbation theory in the transparency region (or static perturbation theory), these are

$$\alpha^{(1)} \equiv \alpha = 2\sum_n{}' \frac{\mu_{gn}\mu_{ng}}{E_{ng}} \tag{2}$$

$$\alpha^{(2)} \equiv \beta = 3\left(\sum_{n,n'}{}' \frac{\mu_{gn}\mu_{nn'}\mu_{n'g}}{E_{ng}E_{n'g}} - \mu_{gg}\sum_n{}' \frac{\mu_{gn}\mu_{ng}}{E_{ng}^2} \right) \tag{3}$$

$$\alpha^{(3)} \equiv \gamma = 4\left(\sum_{n,n'n''}{}' \frac{\mu_{gn}\mu_{nn'}\mu_{n'n''}\mu_{n''g}}{E_{ng}E_{n'g}E_{n''g}} - \sum_n{}' \frac{\mu_{gn}\mu_{ng}}{E_{ng}}\sum_{n'}{}' \frac{\mu_{gn'}\mu_{n'g}}{E_{n'g}^2} \right) \tag{4}$$

where $\sum_m'$ means that terms with $E_{mg} = 0$ will be excluded, m labels the electronic states of the molecule (g is the ground state), and we limit ourselves to the x components; μ is the electronic dipole moment operator. The calculation of these expressions is quite involved, but useful information about size effects and scaling laws can be obtained with the help of the Unsöld approximation (Flytzanis, 1975).

For extended periodic systems, the relevant quantity is now the induced dipole density or polarization P, whose dependence on the applied macroscopic field is expressed in terms of the nonlinear susceptibilities $\chi^{(1)}, \chi^{(2)}, \chi^{(3)}$, etc., and the electron states are represented in terms of Bloch band states. For the derivation of the scaling laws for extended systems that can be related to the ones for small-size systems, the choice of the appropriate expressions of $\chi^{(n)}$ is crucial. This point has been mostly tackled using the Genkin–Mednis approach (Genkin and Mednis, 1968; Flytzanis, 1975), and one obtains

$$\chi_{xx}^{(1)} = \frac{4e^2}{\hbar V} \int_{B.Z.} \Omega_{vc} S_{cv}\, dk \tag{5}$$

$$\chi_{xxxx}^{(3)} = \frac{8e^4}{\hbar^3 V} \int_{B.Z.} \left[\frac{1}{\omega_{vc}} \left(\frac{\partial S_{cv}}{\partial k} \right)\left(\frac{\partial S_{vc}}{\partial k} \right) - \Omega_{vc}S_{cv}S_{vc}S_{cv} \right] dk \tag{6}$$

where $\hbar\omega_{cv} = \varepsilon_c - \varepsilon_v$, Ω_{vc} is the transition dipole-moment matrix element between the highest valence (v) and lowest conduction (c) bands, $S_{vc} = \Omega_{vc}/\omega_{vc}$, and ε_v and ε_c are the band energies for the v and c band, respectively; in Eqs. (5) and (6) we have introduced the two-band approximation, which is quite sufficient for our purpose. Furthermore, we have assumed that the system possesses inversion symmetry so that in particular $\chi^{(2)} \equiv 0$; for noncentrosymmetric systems, the expression of $\chi^{(3)}$ is slightly more lengthy (Agrawal *et al.*, 1978) and the second-order susceptibility is

$$\chi^{(2)} = \left(\frac{6e^3}{\hbar^2 V}\right)\frac{i}{2\pi}\int_{B.Z.}\left[S_{cv}S_{cv}(\Omega_{vv} - \Omega_{cc}) - \frac{1}{2}\left(S_{vc}\frac{\partial S_{cv}}{\partial k} - \frac{\partial S_{vc}}{\partial k}S_{cv}\right)\right]dk \quad (7)$$

where Ω_{vv} and Ω_{cc} are intraband transition dipole-moment matrix elements.

One may formally define also polarizabilities $\alpha^{(n)}$ for such infinitely extended systems by the relation

$$\chi^{(n)} = \frac{\alpha^{(n)}}{v} \quad (8)$$

where v is the repeat unit-cell volume. By simple inspection and comparison of expressions (2), (3), and (4) with (5), (7), and (6) respectively, one may easily see the similarity between these two sets of expressions; as a matter of fact, expressions (2)–(4) change over to expressions (5)–(7) for extended systems with localized electron states.

Before we move to derive the scaling laws using the above expressions, we stress here the fact that the behavior of the nonlinear optical susceptibilities $\chi^{(n)}$ is determined by the competition of two terms, *intraband* ones that arise from field mixing of Bloch band states within a band and *interband* ones that arise from mixing of states across the gap (with wave-vector conservation); in contrast, the linear susceptibility $\chi^{(1)}$ only involves interband terms. For highly delocalized systems (strong overlap between wave functions), the quantities ω_{cv} and Ω_{cv} vary strongly over the Brillouin zone and therefore the intraband term in $\chi^{(3)}$ becomes the dominant one.

As can be seen from Eqs. (5), (6), and (7), the susceptibilities are expressed as integrals over the Brillouin zone. The main contribution to these integrals is in general expected to come from some few nonoverlapping critical regions in the joint density of states (see, for instance, Ashcroft and Mermin, 1981); these are points, lines, and surfaces depending on the spatial extension of the electronic density distribution and are defined by the condition

$$\nabla_{\mathbf{k}}\omega_{cv}(\mathbf{k}) = 0 \quad (9)$$

This is indeed the case for $\chi^{(1)}$ (see, for instance, Cardona and Pollack, 1971) and was found also to be the case for odd-order susceptibilities $\chi^{(2n+1)}$, as will be shown. This directly establishes a relation between the optical susceptibilities $\chi^{(2n+1)}$ and the topology and dimensionality of the joint density of states. More importantly, it allows one to express $\chi^{(2n+1)}$ in terms of the values of Ω_{vc} and ω_{cv} at these critical regions; the latter can also be related to the characteristics of the electron distribution. This functional dependence is precisely the origin of the scaling laws to be given below and also strongly reflects the dimensionality of the electron distribution. The critical point analysis of the even-order susceptibilities $\chi^{(2n)}$ is not as straightforward as for the odd-order ones, as will be cursorily exemplified below for $\chi^{(2)}$, and will not be pursued to any extent here.

III. ZERO-DIMENSIONALITY SYSTEM

Under this headline we loosely include isolated molecular systems of finite extension in all directions and of size much less than the wavelength λ. We shall restrict our discussion to linear centrosymmetric conjugated molecules like polyenes etc. The case of a linear conjugated carbon chain without bond alternation will be tackled first, as it affords almost complete analytical solution using a method initially applied for the investigation of the so called Stark ladder level configuration (Wannier, 1962; Fukuyama $et\ al.$, 1973); subsequently, the case of bond alternation will also be considered, but the analysis will rely on numerical results.

A. Carbon Chains without Bond Alternation

We use the Hückel approximation (Murrell, 1963) so that the coefficients c_n of the wave functions $\psi = \Sigma_c c_n \phi_n$ and their eigenvalues ε for a finite conjugated chain containing $2N$ equally spaced carbon atoms, in the presence of a static electric field E, are determined through

$$c_n(-\varepsilon + nw) + \beta(c_{n+1} + c_{n-1}) = 0 \tag{10}$$

with the boundary conditions $c_0 = c_{2N+1} = 0$; in Eq. (10), $w = eaE$, where a is the interatomic distance; β is the resonance (hopping) energy between neighboring carbon atoms; ε is measured with respect to the Coulomb energy of the carbon chain; and the Hückel approximations, $\langle \phi_n | X | \phi_n \rangle = na$ and $\langle \phi_n | \phi_m \rangle = \delta_{nm}$, were used. The solution of Eq. (10) is

$$c_n = AJ_{v-\varepsilon/w}(-2w/\beta) + BY_{n-\varepsilon/w}(-2w/\beta) = 0$$

where J_v and Y_v are the Bessel and von Newman functions, related (Abramovitz and Stegun, 1964) by

$$Y_v(x) \sin \gamma\pi = J(x) \cos v\pi - J_{-v}(x)$$

The boundary conditions $c_0 = c_{2N+1} = 0$ give

$$J_{-\gamma\xi}(2\gamma) Y_{1-\gamma\xi}(2\gamma) - J_{\gamma/\bar{\gamma}-\gamma\xi} Y_{-\gamma\xi}(2\gamma) = 0$$

where $\gamma = -\beta/w$, $\xi = -\delta/\beta$, and $\bar{\gamma} = (N+1)\gamma$.

The roots of this equation give the $2N$ eigenvalues ε_n, which for the case of $E = 0$ reduce to $\varepsilon_n^0 = 2\beta \cos \theta_n^0$ with $\theta_n^0 = n\pi/(N+1)$, while at the high field intensity limit ($\gamma \ll 1$) one obtains the Stark ladder spectrum $\varepsilon_n = nw$ (Wannier, 1962), and the same is true for intermediate field intensities $\bar{\gamma} \ll 1$ but $\gamma \gg 1$. For low field intensities where perturbation theory can be used — namely, $\gamma \gg 1$ and $\bar{\gamma} \gg 1$ — using the double asymptotic development of Bessel functions (Abramovitz and Stegun, 1964) and rearranging the expressions one obtains for the total energy of the electron system (Agrawal et $al.$, 1978)

$$W = \sum_{n=1}^{N} \varepsilon_n = W_0 - N\beta \sum_{k=0}^{\infty} \frac{\lambda^{(k)}}{\bar{\gamma}^{2k}} \tag{11}$$

where the $\lambda^{(k)}$ are constants that in principle can be calculated. Inserting $\bar{\gamma} = -(N+1)\beta/eaE$, using the definition of the polarizabilities for a symmetric molecule,

$$W = W_0 - \sum_{k=1}^{\infty} \frac{1}{2k-1} \alpha^{(2k-1)} E^{2k-1} \tag{12}$$

and identifying terms of the same order in E in Eqs. (11) and (12), respectively, one obtains (Agrawal et $al.$, 1978).

$$\alpha^{(2n-1)} \approx N^{2n+1} e^{2n} a^{2n}/\beta^{2n-1} = L^{2n+1} e^{2n}/a\beta^{2n-1} \tag{13}$$

valid for large N, and $L = Na$ is the half-length of the chain. In particular for the linear and third order polarizabilities one has

$$\alpha \approx N^3 e^2 a^2/\beta = L^3 e^2/a\beta \tag{14}$$

$$\gamma \approx N^5 e^4 a^4/\beta^3 = L^5 e^4/a\beta^3 \tag{15}$$

These expressions clearly show how the odd-order polarizabilities for linear molecules with equidistant atoms scale with the molecular extension L and resonance energy β and they constitute scaling laws; expression (14) was initially derived by Davies (1952) and expression (15) by Rustagi and Ducuing (1972) using the free electron model to describe such chains (Kuhn, 1948).

These scaling laws should in principle apply for cyanine dyes; however, it is difficult to rigorously test the fifth-power dependence of γ on N. Indeed, as N increases, so does the wavelength of the first absorption peak (Murrell, 1963), and the frequency dependence of γ cannot be overlooked; it may mask any N dependence (Hermann and Ducuing, 1974).

B. Chains with Bond Alternation

These chains are characterized by two resonance energies, β_1 and β_2, and the situation is drastically different from the previous case as bond alternation stabilizes the main absorption peak to a fixed value $2|\beta_1 - \beta_2|$ as the chain length increases (Murrell, 1963); on the other hand, the analytical treatment of the polarizabilities becomes prohibitingly difficult, and meaningful trends can only be extracted numerically. These can be cast in a very simple form if we introduce the delocalization parameter

$$N_d = \frac{\beta_1 + \beta_2}{\beta_1 - \beta_2} \tag{16}$$

We present next the main conclusions regarding the third-order polarizability γ_{2N} for a chain of $2N$ atoms when plotted as a function of N_d for different values of N (Agrawal and Flytzanis, 1976).

For a given N, γ_{2N} shows a maximum at $N_d = 2N$; for $N_d > N$, γ_{2N} decreases monotonically with increasing N_d (and N fixed) and for $N_d \to \infty$ (or $\beta_1 = \beta_2$) one recovers expression (15), namely, a fifth-power dependence of γ_{2N} on N.

For $N_d < N$, γ_{2N} decreases monotomically as

$$\gamma_{2N} \sim N_d^6 \tag{17}$$

independently of N for N_d not too close to unity. For $N_d \approx 1$, or $\beta_2 \approx 0$, γ becomes negative and we reach the expression $\gamma_0 = -Ne^4 a^4/\beta_1^3$ for $\beta_2 = 0$, which corresponds to the value of γ_{2N} for N independent bonds represented by two-level systems; this limiting behavior (bond additivity) clearly must be modified, since the chain group contributions must also be included now and may even dominate the actual value of γ.

The above trend is a clear example of size effect and indicates that the third-order polarizability γ exhibits distinctly different scaling laws depending on whether the electron delocalization length $L_d = N_d a$ is smaller or larger than molecular length $L = Na$; in particular, we expect that the scaling law [Eq. (17)] should be independent of N as long as $N_d < N$, and this is strikingly corroborated by the analytical results of infinite bond-alternated chains (Agrawal and Flytzanis, 1976, and see below).

There has not been any experimental confirmation of the above results for finite chains yet.

IV. ONE-DIMENSIONAL SYSTEMS

We consider now the case of infinite one-dimensional crystals, which can be either one-dimensional organic semiconductors (like polyacetylene or polydiacetylene) or inorganic ones (like SbSI). To gain insight into the scaling laws that prevail here, we exemplify this class of systems with the infinite simply bond-alternated carbon chain and adopt the Hückel approximation to describe the electronic density; within the Hückel approximation two resonance integrals β_1 and β_2 are introduced to characterize the bond alternation as in Section III.B.

It turns out that the scaling laws for $\chi^{(2n+1)}$ derived for this simple system are quite universal and valid for all one-dimensional systems and can be cast in the form of power laws of the electron delocalization length $L_d = N_d a$ where N_d is given by Eq. (16) (Agrawal and Flytzanis, 1976; Agrawal et al., 1978); for the case of $\chi^{(3)}$ it reduces to a sixth-power dependence on N_d, the same one found numerically for finite chains when $N > N_d$ [Eq. (17)].

We assume complete separation between σ and π electrons and we concentrate only on the π-electron contribution. The electron states for such bond-alternated chain can be obtained analytically in the form of Bloch band states. One has in particular

$$\hbar\omega_{cv} = 2\beta_2\sqrt{(1 + v^2 + 2v\cos ka)} \equiv 2\beta_2\zeta_0 \tag{18}$$

$$e\Omega_{cv} = ea(1 - v^2)/4\zeta_0^2 \tag{19}$$

where $v = \beta_1/\beta_2$. Inserting the Eq. (18) for $\hbar\omega_{cv}$ in Eq. (9), we see that the joint density of states becomes infinite at the edge of the Brillouin zone, $ka = \pi$, and this constitutes the critical point where also the matrix element $\Omega_{cv}(ka)$ attains its maximum values,

$$|\Omega_{cv}(\pi)| = a\left|\frac{\beta_1 + \beta_2}{\beta_1 - \beta_2}\right| = L_d \tag{20}$$

which is the optical delocalization length, and the energy difference $\hbar\omega_{cv}$ becomes smallest,

$$\hbar\omega_{cv}(\pi) = 2|\beta_2 - \beta_1| = E_0 \tag{21}$$

which also is the optical gap where the main absorption peak is located. The coincidence of these three features at a single point of the Brillouin zone— namely, (1) infinite joint density of states, (2) maximum transition dipole moment, and (3) smallest energy gap—is an essential characteristic of the one-dimensional semiconductors, reminiscent of a metallic-type behavior (this point is the only one preserved from the metallic case when bond alternation sets in), and is the origin of the enhanced contribution of the π electrons over the σ electrons in the optical susceptibilities. Indeed, inserting Eqs. (18) and (19) in Eqs. (5) and (6) one finds for strong electron delocalization, $N_d = 4L_d/a > 1$,

$$\chi_\pi^{(1)} = \frac{2}{3\pi}\chi_\sigma^{(1)}N_d^2 \tag{22}$$

$$\chi_\pi^{(3)} = \frac{2}{3\pi}\chi_\sigma^{(3)}N_\sigma^6 \tag{23}$$

and quite generally one can show that

$$\chi_\pi^{(2n-1)} \approx \chi_\sigma^{(2n-1)} N_d^{4n-2} \tag{24}$$

where $\chi_\sigma^{(2n-1)}$ is the expression of the susceptibility for a chain of saturated bonds (σ electrons), which corresponds to setting $\beta_2 = 0$ in Eqs. (18) and (19) or

$$\chi_\sigma^{(2n-1)} \sim (ea)^{2n}/\beta_2^{2n-1} \tag{25}$$

Expression (23) and the more general one of Eq. (24) are valid for a large electron delocalization $N_d > 1$; it was found that the main contribution in the odd-order nonlinear susceptibilities comes from the intraband terms, which have opposite sign to the interband ones (Agrawal *et al.*, 1978). Since $N_d > 1$, comparison of Eqs. (24) and (25) shows that indeed the nonlinear susceptibilities for π-electron systems are greatly enhanced by many orders of magnitude over those of the σ electrons (for the same electron density), and in particular this justifies the neglect of the side-group contributions in these chains, since these usually are saturated molecular systems.

The detailed analysis of Agrawal *et al.* (1978) showed that the scaling law of Eq. (24) is not restricted to simply bond-alternated chains but is valid for all one-dimensional systems, with or without inversion symmetry, bond super-alternation, coupled chains, etc., and similarly one expects that the scaling law of Eq. (25) for higher-order susceptibilities is universally valid for all one-dimensional systems with delocalized electron states (Flytzanis, 1983).

Clearly the universality of the scaling law of Eq. (24) or the more general one of Eq. (25) has only been proven within the one-electron approach where electron correlation is altogether neglected (Ovchinnikov, 1973); when the latter is taken into account, the expressions of Ω_{cv} and $\chi^{(n)}$ become too exceedingly complex to allow any analytical or even numerical treatment, and no conclusions can be drawn yet.

In practice, one-dimensional conjugated chains are seldom of infinite extension; they are more usually interrupted by different types of defects, extrinsic (impurities etc.) or intrinsic (conjugation or Pople–Wamsley defects, solitons, etc). The question arises then as to what extent the scaling laws of Eq. (24) or (25) obtained for an infinite chain are still valid for chains formed by segments of $2N$ atoms. For $\chi^{(3)}$, the answer is contained in Section III,B, where it was found that indeed the third-order polarizability of such a segment still exhibits the sixth-power dependence on N_d as long as $N_d > 1$, and this independently of N as long as $N_d < N$; thus the scaling law of Eq. (24) is still valid, and similarly we expect this to be the case for the scaling law of Eq. (25).

For the subsequent discussion it is convenient to cast the scaling law of Eq. (25) in a form more closely related to solid-state terminology. Indeed, we note that for the simply bond-alternated chain, when considered as a one-dimensional semiconductor, the Fermi energy $E_F = |\beta_2 + \beta_1|$ and the energy gap $E_0 = 2|\beta_1 + \beta_2|$ so that $N_d \approx E_F/E_0$ or

$$\chi^{(3)} \approx \chi_\sigma^{(3)} \left(\frac{E_F}{E_0}\right)^6 \tag{26}$$

and more generally

$$\chi^{(2n-1)} \approx \chi_\sigma^{(2n-1)} \left(\frac{E_F}{E_0}\right)^{4n-2} \tag{27}$$

which contains quantities directly accessible to measurement, the Fermi energy and the energy gap. This form in particular allows one to compare organic with inorganic semiconductors (like SbSI), and more importantly it directly relates $\chi^{(2n-1)}$ to the critical-point energy E_0; the sixth-power dependence of $\chi^{(3)}$ on E_F/E_0 was first derived by Cardona and Pollack (1971) using critical-point analysis and also by Sauteret *et al.* (1976) using a simple model of a one-dimensional semiconductor.

Before moving into the case of higher-dimensionality systems, we wish to say a few words about the second-order susceptibility $\chi^{(2)}$, given by Eq. (7). Here the interband and intraband terms can each have either sign, so that they may add or subtract. Careful analysis of Eq. (7) reveals that contributions to the integrand only come from regions were Ω_{cv} is complex (Agrawal *et al.*, 1978); it vanishes whenever Ω_{vc} becomes real or pure imaginary, and this is precisely what happens at the edge of the Brillouin zone of a one-dimensional system. Thus, in contrast to $\chi^{(3)}$, $\chi^{(2)}$ does not take full advantage of the infinite densities of states there and the highly delocalized character of these states; in particular, the critical-point analysis for the derivation of scaling laws is not as straightforward as for $\chi^{(3)}$.

V. TWO- AND THREE-DIMENSIONAL SYSTEMS

For two- and three-dimensional crystalline systems, the electronic band states and energies cannot be obtained easily in a closed analytical form over the whole Brillouin zone, and the calculation of the susceptibilities using Eqs. (5–7) can only be conducted numerically. However as pointed previously, the main contribution to the integrals in expressions (5–7) comes from some few nonoverlapping critical regions in the joint density of states, and one then may use the approach of Cardona and Pollack (1971) based on a model density of states to obtain the contributions to $\chi^{(2n-1)}$ from the different critical points.

A. Two-Dimensional Systems

Here the critical regions are of two kinds: a *point*, at energy E_0 at the edge of the Brillouin zone, reminiscent of one-dimensional patterns in the electron density distribution, with their contribution to $\chi^{(2n-1)}$ given by Eq. (27); and a *line*, at energy E_1, intrinsic to the two-dimensional character of the system. In general, $E_1 < E_0$, and furthermore the impact of E_0 is now reduced as E_1 attracts a large density of states. Cardona and Pollack (1971), using plausible simplifications in their model density of states, show that the contribution to $\chi^{(1)}$ and $\chi^{(3)}$ from the critical points on E_1 are

$$\chi^{(1)} \sim P^2/E_1^2 \tag{28}$$

$$\chi^{(3)} \sim P^2/E_1^5 \tag{29}$$

respectively, and by induction we make the conjecture that

$$\chi^{(2n-1)} \sim P_0^2/E_1^{3n-1} \tag{30}$$

where P is an average transition dipole moment matrix element.

As in the one-dimensional semiconductor case, the scaling law of Eq. (29) arises from competition of field mixing of states within each band (intraband terms) and k-conserving coupling across the gap (interband terms). This competition can also be interpreted as displacements of opposite sign of the band gap at the critical points when an electric field is applied (Cardona and Pollack, 1971).

B. Three-Dimensional Systems

Here, besides contributions from critical points E_0 and lines E_1 of the form of Eqs. (27) and (30), respectively, $\chi^{(2n-1)}$ will also contain contributions from critical surfaces at energy E_2, and in general $E_2 < E_1 < E_0$, so that for a given electron density the impact of E_0 is further reduced.

In particular, since ω_{cv} and Ω_{cv} vary much more rapidly as we move from E_0 to E_2, the corresponding intraband terms also become more dominant.

Again using simple considerations for the model density of states, one obtains (Cardona and Pollack, 1971)

$$\chi^{(1)} = P^2/E_2^{5/2} \tag{31}$$

$$\chi^{(3)} = P^2/E_2^{9/2} \tag{32}$$

and by induction we make the conjecture that

$$\chi^{(2n-1)} \sim P^2/E_2^{2n+1/2} \tag{33}$$

Clearly these scaling laws are subject to large uncertainties, as they were derived from a very approximate model for the density of states.

VI. MICROSTRUCTURES AND CONCLUSION

The previous scaling laws were derived for systems with infinite extension in one, two, or three dimensions. For many important applications of nonlinear optics, however, systems with reduced extension will increasingly be used; accordingly, size effects will be crucial and some modifications must be introduced. This has been extensively discussed for the one-dimensional systems, where it was shown (Sections III and IV) that as long as the electron delocalization length L_d is smaller than the chain length L, the scaling laws have the same form as for the infinite one-dimensional systems given by Eqs. (23) and (24).

For two- and three-dimensional systems, the validity range of the scaling laws of Eqs. (29) and (32) or the more general ones of Eqs. (30) and (33), respectively, cannot be obtained as easily, and the whole problem has to be reconsidered carefully.

In addition, we may also consider the case where systems of finite extension are imbedded in a different medium of infinite extension, usually amorphous or glass; these are the so-called composite materials, like metal inclusions in glasses (Peremboom and Wyder, 1981) or semiconductor-doped glasses. Here, local field effects mediated through the surrounding medium can play a crucial role under certain resonance conditions and lead to an enhancement of the nonlinear susceptibilities. The effective medium approach (Maxwell–Garnett theory) has been extended to the nonlinear case to take into account these effects (Rustagi and Flytzanis, 1984; Ricard *et al.*, 1985), but more work must be done on these systems to clearly understand the origin of the nonlinear response and the scaling laws that prevail there. More recently, Hache *et al.* (1986) have discussed quantum size effects in the third-order polarizability of small metallic particles and derived the expression of the third-order susceptibility of the corresponding composite material; the dielectric confinement of the electrons in these metallic particles leads to a strong dependence of the magnitude of $\chi^{(3)}$ on the size of the particles. The case of the semiconductor-doped glasses provides another class of composite materials where size effects play an important role in the nonlinear optical properties. Here, too, the dielectric confinement drastically modifies the nature of the electronic states, the oscillator strength as well as the response time of the nonlinearity [Roussignol *et al.* (1986)]. This in particular leads to an intensity dependence, of the optical Kerr effect and the optical-phase-conjugated reflectivity in these materials, that was termed "frustated saturation". [Roussignol *et al.* (1986)].

The study of the nonlinear optical properties of composite materials is interesting in a number of other aspects as it is related to that of small aggregates that are of paramount importance in many areas in chemistry and metallurgy. From a more fundamental point of view, the physics of such a low-dimensionality system throws a new light on some basic physical principles, which are suppressed when we deal with infinitely large systems. These heterogeneous materials present a lot of advantages when compared to the homogeneous ones, since one can artificially change a number of characteristics, including the size of inclusions, their resonances or band gap, their concentration, and even the optical anisotropy by using inclusions of appropriate shape, and thus manufacture materials with "tunable nonlinearity," which are of great importance in many applications of nonlinear optics [Flytzanis *et al.* (1986)].

As stated in the introduction, scaling laws can be very useful and in many considerations essential for our understanding of the physics of the nonlinearities. The approach presented above is only tentative, as experimental investigations are totally lacking. However, with the increasing use of microstructures and the requirements that prevail there, dimensionality considerations will play a very important role, and so will the scaling laws.

REFERENCES

Abramovitz, M., and Stegun I. (1964). "Handbook of Mathematical Function." Dover, New York.

Agrawal, G. P., and Flytzanis, C. (1976). *Chem. Phys. Lett.* **44**, 366.

Agrawal, G. P., Cojan, C., and Flytzanis, C. (1978). *Phys. Rev. B* **17**, 776.

Ashcroft, N. W., and Mermin, N. D. (1981). "Solid State Physics." Holt Saunders, Tokyo.

Cardona, M., and Pollack, F. H. (1971). *In* "Optoelectronic Materials" (G. A. Albers, ed.), Plenum, New York.

Cojan, C., Agrawal, G. P., and Flytzanis, C. (1977). *Phys. Rev. B* **15**, 909'

Davies, P. L. (1952). *Trans. Faraday Soc.* **48**, 789.

Flytzanis, C. (1975). *In* "Treatise of Quantum Electronics" (H. Rabin and C. L. Tang, eds.), Vol. 1A, Part A. Academic Press, New York.

Flytzanis, C. (1983). *ACS Symp. Ser.* **233**, 167.

Flytzanis, C., and Ducuing, J. (1969). *Phys. Rev.* **178**, 1218.

Flytzanis, C., Hache, F., Ricard, D., and Roussignol, P. (1986). *In* "The Physics and Fabrication of Microstructures and Microdevices" (M. J. Kelly and C. Weisbuch, eds.), p. 331. Springer–Verlag, Berlin and New York (in press).

Fukuyama, H. Bari, R. A., and Fogedby, B. (1973). *Phys. Rev. B* **8**, 5579.

Genkin, V. N., and Mednis, P. M. (1968). *Sov. Phys.—JETP (Engl. Transl.)* **27**, 609; *Zh. Eksp. Teor. Fiz.* **54**, 1137.

Hache, F. Ricard, D., Flytzanis, C. (1986). *J. Opt. Soc. Am. B* (to be published).

Hermann, J. P., and Ducuing, J. (1974). *J. Appl. Phys.* **45**, 5100.

Kuhn, H. (1948). *J. Chem. Phys.* **16,** 840.

Miller, R. C. (1964). *Appl. Phys. Lett.* **50,** 17.

Murrell, J. N. (1963). "The Theory of the Electronic Spectra of Organic Molecules." Methuen, London.

Ovchinnikov, A. A. (1973). *Sov. Phys.—Usp. (Engl. Transl.)* **15,** 575.

Perenboom, J. A. A. J., and Wyder, P. (1981). *Phys. Rep.* **78,** 173.

Ricard, D., Roussignol, P., and Flytzanis, C. (1985). *Opt. Lett.* **10,** 511.

Roussignol P., Ricard, D., Lukasik, J., Flytzanis, C. (1986). *J. Opt. Soc. Am. B* (to be published).

Rustagi, K., and Ducuing, J. (1972). *Opt. Commun.* **10,** 258.

Rustagi, K., and Flytzanis, C. (1984). *Opt. Lett.* **9,** 344.

van Vechten, J. A., Cardona, M., Aspnes, D. A., and Martin, R. M. (1970). *Proc. Int. Conf. Semicond.*, 1970, p. 82.

Wannier, G. (1962). *Rev. Mod. Phys.* **34,** 645.

Chapter III-5

Trends in Calculations of Polarizabilities and Hyperpolarizabilities of Long Molecules

JEAN-MARIE ANDRÉ, CHRISTIAN BARBIER,
VINCENT BODART, and JOSEPH DELHALLE

Laboratoire de Chimie Théorique Appliquée
Facultés Universitaires Notre-Dame de la Paix
Rue de Bruxelles, 61, B-5000 Namur, Belgium

I. INTRODUCTION

In this chapter, we report on some aspects of *ab initio* computations of electric polarizabilities and hyperpolarizabilities of relatively large molecules, and consider, as an illustration, the influence of chain length and bond alternation on these properties in the polyene series $H—(CH=CH)_x—H$ for $x = 1, 2, 3,$ and 4.

In the first part and in the line of our theoretical work on organic chain molecules and polymers, we focus our attention on *ab initio* calculations

137

carried out within the framework of the Hartree–Fock approximation. We investigate, at this level of theory, the use of finite-field (FF) and perturbative (SOS, summation over states) methods. The size dependence is analyzed and compared to previously published results using simpler models (free-electron, Hückel, and extended Hückel).

In a second part, we discuss the question of the reference state used in perturbation approaches and stress that a proper comparison between FF and SOS results should imply double perturbation series in order to take into account the effects of electronic rearrangments in presence of external field.

Finally, in the last part, we comment on the problem of extrapolating results obtained on chains of limited size to large polymeric compounds.

II. FF AND SOS STUDIES OF (HYPER)POLARIZABILITIES OF SMALL-CHAIN MOLECULES

In this part we begin with a brief overview of both the FF and SOS methods, and then report on the dependence of (hyper)polarizabilities on geometrical changes for small polyene chains as predicted by both approaches.

A. The RHFR Method

The reference methodology in this chapter is the restricted Hartree–Fock Roothan (RHFR) model (Roothan, 1951). At this level, one considers the independent motion of a single electron in the electrostatic field of the nuclei and in the average Coulomb and exchange fields of all other electrons. That level of the theory results in the traditional molecular orbital (MO) language. The wave function is taken as an antisymmetrized product of one-electron orbitals, which are self-consistent eigenfunctions of the one-electron Fock operator, $h_0(r)$. This operator contains kinetic, electron–nuclei attraction, electron–electron Coulomb repulsion, and electron–electron Pauli exchange interaction parts. In the LCAO (linear combination of atomic orbitals) scheme, the one-electron Hamiltonian is represented by its matrix elements over a truncated set of N atomic orbitals:

$$\langle p|F_0|q\rangle = \langle p|-\tfrac{1}{2}\nabla^2|q\rangle - \sum_A^{N_A}\left\langle p\left|\frac{Z_A}{R_A}\right|q\right\rangle$$
$$+ \sum_i\sum_{rs} C_{ir}C_{is}\{(pq|rs) - \tfrac{1}{2}(pr|qs)\} \tag{1}$$

The expression is given for a molecule containing N_A nuclei (each one of charge Z_A) and $2K$ electrons. The one-electron integrals (kinetic and nuclear attraction)

$$\langle p|G|q\rangle = \int dv\, \chi_p(r)G(r)\chi_q(r) \tag{2}$$

form a square matrix of order N whose only $M = N(N + 1)/2$ nonredundant elements are calculated. The bottleneck of the RHFR procedure is the calculation of the N^4 two-electron integrals:

$$(pq|rs) = \int\int dv_1\, dv_2\, \chi_p(1)\chi_q(1)r_{12}^{-1}\chi_r(2)\chi_s(2) \tag{3}$$

Even if only the $M(M + 1)/2$ nonredundant terms are evaluated, this part needs careful optimization in order to be efficient in actual calculations (Pople and Hehre, 1978). The summation over the K doubly occupied orbitals ensures the self-consistency of the procedure and defines the elements of a density matrix:

$$D_{rs} = \sum_i^K C_{ir}C_{is} \tag{4}$$

The calculations described in this chapter have been performed at three degrees of sophistication in the extent of the LCAO expansion of the molecular orbitals; the STO-3G (Hehre $et\ al.$, 1969), 4-31G (Ditchfield $et\ al.$, 1971), and 6-31G** (Hariharan and Pople, 1973) basis sets. In the STO-3G calculations, the molecular orbitals are expanded in the occupied atomic orbitals (AO) of the isolated atoms, i.e., 1s for hydrogen, 1s, 2s, $2p_x$, $2p_y$, $2p_z$ for carbon. In this method, each Slater-type orbital (STO) is a linear combination of three gaussians. Their parameters (coefficients and exponents of the gaussian expansion) have been optimized to reproduce selected molecular properties such as molecular geometry. For qualitative interpretations, minimal basis sets have the further great advantage of relating molecular properties to simple atomic parameters and allow a way of thinking common to both theoreticians and experimentalists. In order to provide a better flexibility to the basis set and hence a better description of the electron density, it is recommended to add to each valence atomic orbital a more delocalized function to reproduce the effects of bonding in the interatomic regions. In the 4-31G basis set, core orbitals (1s of C) are expanded in four gaussians, while valence orbitals (2s, $2p_x$, $2p_y$, $2p_z$ of C, 1s of H) are described by two linearly independent basis functions consisting, respectively, of three gaussians and a single gaussian. Further improvements along these lines can be obtained with the 6-31G** basis, which differs qualitatively from the 4-31G basis in only two respects. First, the innermost shell (i.e., the core orbitals) is a contraction of six gaussians instead of four; and second, a d-shell for carbon and a p-shell for hydrogen are added.

B. Basic Equations of the FF and SOS Methods

In the framework of the Bloembergen model (Bloembergen, 1965), and considering homogeneous electric fields, the dipole moment of the molecule, μ_i, is represented as a power series expansion in terms of the electric field strengths F_i,

$$\mu_i = \mu_{i0} + \alpha_{ij}F_j + \beta_{ijk}F_jF_k + \gamma_{ijkl}F_jF_kF_l + \cdots \tag{5}$$

where μ_{i0}, α_{ij}, β_{ijk}, and γ_{ijkl}, respectively, are the components of the permanent dipole μ_0, the linear α, second-order β, and third-order γ, polarizabilities that we have to compute.

Since the finite-field method has already been described in this book and in the literature, we shall limit ourselves to the basic aspects of the theory. We shall insist later on the numerical implications of finite-field calculations. The FF—SCF (finite field—self-consistent field) method was originally proposed by Cohen and Roothan (1965). It is equivalent to an analytic coupled Hartree–Fock scheme. A term, μF, describing the interaction between the external field F and the elementary charges (electrons and nuclei) constituting the molecule, is added to the molecular Hamiltonian; μ is the total dipole moment of the molecule. At the RHF level, the one-electron orbitals are self-consistent eigenfunctions of a one-electron field-dependent Fock operator, $h(r)$:

$$h(r) = h_0(r) + er \cdot F \tag{6}$$

The matrix elements of $h(r)$ contain additional one-electron moment integrals, $\langle p|r|q \rangle$, whose calculation is fairly standard. However, one has to realize that the self-consisent procedure applies to the solutions of $h(r)$. More specifically, the elements D_{pq} of the density matrix are now field-dependent, $D_{pq}(F)$. The numerical approach to the calculation of electronic (hyper)polarizabilities requires much higher accuracy for the total SCF energy than in standard applications. The convergence threshold on the density matrix elements has to be as small as 10^{-9}. The components of the polarizability tensor α_{ii} ($i = x, y, z$) are obtained from the first derivative of the field-dependent dipole moment, $\mu(F)$, with respect to F_i in the limit of zero field. In practice, this is done numerically using

$$\alpha_{ii} = [\mu_i(F_i) - \mu_i(-F_i)]/2F_i \tag{7}$$

The FF polarizability calculations reported in this paper were carried out with four values of the external field: -0.002, -0.001, 0.001, and 0.002 a.u. (1 a.u. $= 1.72 \times 10^5$ esu). The Richardson extrapolation procedure (Rutishauser, 1963) was applied to check the numerical stability of the results, a point that remains difficult in the finite-field approach. The computation of the second hyperpolarizability tensor (γ) can in principle be done according to the same scheme, the third derivatives being evaluated using the following formula:

$$\gamma_{iiii} = \{\mu_i(3F_i) - \mu_i(-3F_i) - 3[\mu_i(F_i) - \mu_i(-F_i)]\}/48F_i^3 \tag{8}$$

In practice, however, the number of field values at which the induced dipole has to be calculated turns out to be much larger than in the case of the polarizability tensor. Moreover, the values of the field that should ideally be used in these calculations are such that numerical stability cannot be reached to within the required accuracy threshold. Owing to this, we are not in a position to report here on the components of the second hyperpolarizability tensor calculated with the FF method.

Summarizing, the RHF-FF-SCF method implies not only the (almost non-time-consuming) calculation of dipole integrals $\langle p|r|q \rangle$ but also, for each value of the external electric field that is needed in the differentiation procedure, a complete SCF set of iterations. This becomes computationally heavy and uncertain when the second hyperpolarizability γ is considered.

On the other side, the SOS method (e.g., Szabo and Ostlund, 1982) does not involve a new set of SCF procedures but instead uses as a starting point the RHF wave function (or the corresponding SCF-MO orbitals) in the *absence* of the external field. The RHF wave functions are considered as zero-order eigenfunctions of an (undefined) unperturbed hamiltonian with an unperturbed zero-order energy $E^{(0)}$, which is a sum of one-electron energies $\varepsilon_a^{(0)}$:

$$E^{(0)} = \sum_a \varepsilon_a^{(0)} \tag{9}$$

The nature of the reference unperturbed state is not a trivial point and will be discussed in the second part of this chapter. A power-series expansion of the perturbed ground-state energy E_0 is calculated by time-independent perturbation theory with the previously defined perturbation $-\mu F$. The polarizability tensor and higher-order hyperpolarizability tensors can be evaluated from the power-series expansion of the dipole moment in terms of the electric field and from the use of the Hellmann–Feynman theorem (see, for instance, Atkins, 1983):

$$\mu_i^{(0)} = -\left.\frac{\partial E_0}{\partial F_i}\right|_{F=0} \tag{10}$$

In principle, the various energy corrections can be evaluated to arbitrary order using standard time-independent perturbation theory. Diagrammatic techniques can be very helpful, at least to check the rather involved analytical expressions that emerge when orders as high as the fourth (necessary for the actual calculation of the second hyperpolarizability γ) are needed. Following the lines of Szabo and Ostlund (1982), we present in Fig. 1 the Hugenholtz diagrams that represent the perturbation expansion for the ground-state energy up to the fourth order. From the graphs and the standard rules for evaluating the corresponding perturbation expansion, we obtain the expressions for the energy corrections up to the fourth order:

$$E_0^{(0)} = \sum_a \varepsilon_a^{(0)} \tag{11}$$

$$E_0^{(1)} = \sum_a v_{aa} \tag{12}$$

$$E_0^{(2)} = \sum_{ar} \frac{v_{ra} v_{ar}}{\varepsilon_a^{(0)} - \varepsilon_r^{(0)}} \tag{13}$$

$$E_0^{(3)} = \sum_{ars} \frac{v_{ra} v_{as} v_{sr}}{(\varepsilon_a^{(0)} - \varepsilon_r^{(0)})(\varepsilon_a^{(0)} - \varepsilon_s^{(0)})} - \sum_{abr} \frac{v_{ra} v_{ab} v_{br}}{(\varepsilon_a^{(0)} - \varepsilon_r^{(0)})(\varepsilon_b^{(0)} - \varepsilon_r^{(0)})} \tag{14}$$

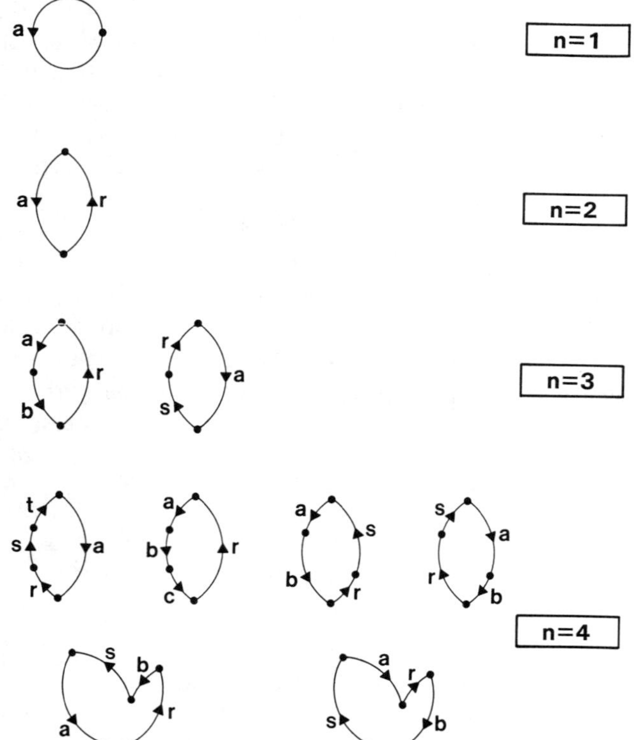

Fig. 1. One-electron Hugenholtz perturbation diagrams up to fourth order.

$$E_0^{(4)} = \sum_{arst} \frac{v_{ar}v_{rs}v_{st}v_{ta}}{(\varepsilon_a^{(0)} - \varepsilon_r^{(0)})(\varepsilon_a^{(0)} - \varepsilon_s^{(0)})(\varepsilon_a^{(0)} - \varepsilon_t^{(0)})}$$

$$+ \sum_{abcr} \frac{v_{ab}v_{bc}v_{cr}v_{ra}}{(\varepsilon_a^{(0)} - \varepsilon_r^{(0)})(\varepsilon_b^{(0)} - \varepsilon_r^{(0)})(\varepsilon_c^{(0)} - \varepsilon_t^{(0)})}$$

$$- \sum_{abrs} \frac{v_{ab}v_{bs}v_{sr}v_{ra}}{(\varepsilon_a^{(0)} - \varepsilon_r^{(0)})(\varepsilon_b^{(0)} - \varepsilon_s^{(0)})(\varepsilon_b^{(0)} - \varepsilon_r^{(0)})}$$

$$- \sum_{abrs} \frac{v_{ab}v_{br}v_{rs}v_{sa}}{(\varepsilon_a^{(0)} - \varepsilon_r^{(0)})(\varepsilon_a^{(0)} - \varepsilon_s^{(0)})(\varepsilon_b^{(0)} - \varepsilon_s^{(0)})}$$

$$- \sum_{abrs} \frac{v_{ar}v_{rb}v_{bs}v_{sa}}{(\varepsilon_a^{(0)} - \varepsilon_r^{(0)})(\varepsilon_a^{(0)} - \varepsilon_s^{(0)})(\varepsilon_b^{(0)} - \varepsilon_s^{(0)})} \tag{15}$$

Setting the perturbation $v = -\mu F$ and proceeding along the lines previously given, we obtain the expressions of the polarizabilities and of the first and second hyperpolarizabilities in the SOS method:

$$\alpha_{ij}^{(0)} = S(ij) \sum_{ar} \frac{\langle a|\mu_i|r\rangle\langle r|\mu_j|a\rangle}{\varepsilon_r^{(0)} - \varepsilon_a^{(0)}} \tag{16}$$

$$\beta_{ijk}^{(0)} = S(ijk) \left[\sum_{ars} \frac{\langle r|\mu_i|a\rangle\langle a|\mu_j|s\rangle\langle s|\mu_k|r\rangle}{(\varepsilon_r^{(0)} - \varepsilon_a^{(0)})(\varepsilon_s^{(0)} - \varepsilon_a^{(0)})} \right. \tag{17}$$

$$\left. - \sum_{abr} \frac{\langle r|\mu_i|a\rangle\langle a|\mu_j|b\rangle\langle b|\mu_k|r\rangle}{(\varepsilon_r^{(0)} - \varepsilon_a^{(0)})(\varepsilon_r^{(0)} - \varepsilon_b^{(0)})} \right]$$

$$\gamma_{ijkl}^{(0)} = S(ijkl) \left[\sum_{arst} \frac{\langle a|\mu_i|r\rangle\langle r|\mu_j|s\rangle\langle s|\mu_k|t\rangle\langle t|\mu_1|a\rangle}{(\varepsilon_r^{(0)} - \varepsilon_a^{(0)})(\varepsilon_s^{(0)} - \varepsilon_a^{(0)})(\varepsilon_t^{(0)} - \varepsilon_a^{(0)})} \right.$$

$$+ \sum_{abcr} \frac{\langle a|\mu_i|b\rangle\langle b|\mu_j|c\rangle\langle c|\mu_k|r\rangle\langle r|\mu_1|a\rangle}{(\varepsilon_r^{(0)} - \varepsilon_a^{(0)})(\varepsilon_r^{(0)} - \varepsilon_b^{(0)})(\varepsilon_r^{(0)} - \varepsilon_c^{(0)})}$$

$$- \sum_{abrs} \frac{\langle a|\mu_i|b\rangle\langle b|\mu_j|s\rangle\langle s|\mu_k|r\rangle\langle r|\mu_1|a\rangle}{(\varepsilon_r^{(0)} - \varepsilon_a^{(0)})(\varepsilon_s^{(0)} - \varepsilon_b^{(0)})(\varepsilon_r^{(0)} - \varepsilon_b^{(0)})}$$

$$- \sum_{abrs} \frac{\langle a|\mu_i|b\rangle\langle b|\mu_j|r\rangle\langle r|\mu_k|s\rangle\langle s|\mu_1|a\rangle}{(\varepsilon_r^{(0)} - \varepsilon_a^{(0)})(\varepsilon_s^{(0)} - \varepsilon_a^{(0)})(\varepsilon_r^{(0)} - \varepsilon_b^{(0)})}$$

$$\left. - \sum_{abrs} \frac{\langle a|\mu_i|r\rangle\langle r|\mu_j|b\rangle\langle b|\mu_k|s\rangle\langle s|\mu_1|a\rangle}{(\varepsilon_r^{(0)} - \varepsilon_a^{(0)})(\varepsilon_s^{(0)} - \varepsilon_a^{(0)})(\varepsilon_s^{(0)} - \varepsilon_b^{(0)})} \right] \tag{18}$$

where $S(ij)$, $S(ijk)$, and $S(ijkl)$ mean symmetrization of the corresponding expressions over indices ij, ijk, and $ijkl$, respectively. All polarizability and hyperpolarizability tensors can be supposed symmetric with respect to all $ijkl$ indices in the static limit. For frequency-dependent polarizabilities and hyperpolarizabilities, however, this symmetry property can only be guaranteed for situations sufficiently far from resonance frequencies (Butcher, 1965). The equations given for α, β, and γ [Eqs. (16)–(18)] can be further simplified in the case of closed-shell structures. The summations over spin orbitals are replaced by summations over spatial orbitals, and the corresponding expressions are multiplied by a factor of 2.

The former expressions are given in terms of integrals over molecular orbitals, and in general involve quadruple summations over molecular orbitals (occupied or unoccupied in the ground state). It should be noted that Hameka has used in his pioneering series of papers a different approach based on molecular eigenstates (McIntyre and Hameka, 1978a–c, 1979; Zamani-Khamiri and Hameka, 1979, 1980; Zamani-Khamiri et al., 1980a,b). He gives the formula for the fourth-order correction to the ground-state energy (McIntyre and Hameka, 1978c) as

$$E_0^{(4)} = - \sum_{k \neq 0} \sum_{L \neq 0} \sum_{M \neq 0} \frac{V_{0K} V_{KL} V_{LM} V_{M0}}{(E_K - E_0)(E_L - E_0)(E_M - E_0)} \tag{19}$$

$$+ \sum_{K \neq 0} \frac{V_{0K} V_{K0}}{(E_K - E_0)} \sum_{L \neq 0} \frac{V_{0L} V_{L0}}{(E_L - E_0)^2}$$

where $|K\rangle$, $|L\rangle$, and $|M\rangle$ are zero-order molecular eigenstates. According to Hameka, the matrix elements are defined as

$$V_{KL} = \langle K|V|L \rangle - \langle 0|V|0 \rangle \delta_{KL} \tag{20}$$

In the original paper by Hameka, the δ_{KL} symbol is missing, which is probably a misprint. It can straightforwardly be seen that the expansion in terms of molecular N-electron eigenstates is equivalent to the previously given expressions over one-electron molecular orbitals [Eqs. (16)–(18)]. Furthermore, the expressions involve a triple summation "only"; due to the one-electron character of the perturbation operator, the V_{0K} interactions are nonzero between singly excited eigenstates; for the same reason, the V_{KL} interactions must be calculated only between a single excited wave function $|K\rangle$ and a doubly excited one $|L\rangle$, which differs from $|K\rangle$ by a single molecular orbital.

Summarizing, the SOS method implies a single SCF calculation without any external perturbating field, contrary to the finite-field theory, which demands several SCF calculations for selected values of the external field. As in the FF approach, the SOS requires the evaluation of the one-electron moment integrals. Using Eq. (16), the calculation of the linear polarizability (α) is quite fast, while it is a much more involved task for γ when Eq. (18) is literaly implemented. Developing efficient algorithms for computing fourth-order corrections is thus a timely endeavor.

C. Application to the Series of Polyenes

Using the aforementioned *ab initio* methods and considering the first terms in the polyene series, effects of chain length and bond alternation on the linear polarizability and the second hyperpolarizability are analyzed. Compounds studied are ethylene, C_2H_4; 1,3-*trans*-butadiene, C_4H_6; 1,3,5-*trans*-hexatriene, C_6H_8; and 1,3,5,7-*trans*-octatetraene, C_8H_{10}.

For the regular polyene chains, both free-electron and Hückel theories predict a behavior of the longitudinal π-electron polarizability like L^3 (where L is the total chain length) and an L^5 dependence for the second hyperpolarizability. Thus, chain length appears, at first sight, as a potentially useful parameter to consider in generating high values of α and γ. Let us note, however, that such an L^3 (or L^5) dependence is only valid in the limit of large N in the free-electron model (Rustagi and Ducuing 1974; Hermann, 1974; Hermann and Ducuing, 1974). The trends are not as well defined in the Hückel methodology. The first published work (Davies, 1952) claims an asymptotic polarizability proportional to the cube of the molecular length, but the perturbation expansion is only approximatively solved. Hameka (1977) started also from the Hückel equation in the presence of an electric field and expanded the equations in term of the electric field by using perturbation theory. He was able to solve the successive perturbation equations analytically and derived exact closed-form expressions for the energy perturbations up to fourth order. From tabulated values of the expansion, he found that the linear polarizability is roughly proportional to the 2.828 power of N and that the nonlinear susceptibility is roughly proportional to the 5.319 power of N.

On different grounds, P. Lambin (unpublished results, 1985) has made an exact derivation of the polarizability in the Hückel framework by means of the resolvant matrix method and a continued fraction expansion. The resulting expression is

$$\alpha = \left(\frac{N^3}{12}\right)\frac{\mathcal{N}(E_F)}{1 + \mathcal{N}(E_F)}(ea)^2 \tag{21}$$

where N is the number of atoms, $\mathcal{N}(E_F)$ the density of states at the Fermi level, a the lattice period, and U an effective intraatomic electron–electron interaction.

The Hückel approach suffers from two drawbacks. First, when strictly applied, it concerns systems with equal bond lengths, while it is well known however that polyenes have a strongly alternant structure. Second, the simple Hückel method does not take into account Coulombic interactions explicitly. It is thus relevant to perform calculations of polarizabilities and hyper-polarizabilities using *ab initio* methods, since a wider spectrum of interesting properties including equilibrium geometries can be consistently evaluated within the limits of a chosen basis set. Up to now, a few FF *ab initio* studies have been devoted to the study of the longitudinal polarizabilities of conjugated hydrocarbon chains (Bodart *et al.*, 1985a,b; Delhalle *et al.*, 1985). Hereafter we take the opportunity to compare some components of the electric polarizability tensor calculated with minimal STO-3G and split-valence 4-31G bases in both the FF and SOS methodologies. The results are listed in Table I and depicted in Fig. 2. The size dependence with respect to N is as follows.

α_{xx}	FF STO-3G	$N^{0.958}$
	SOS STO-3G	$N^{1.010}$
α_{yy}	FF STO-3G	$N^{0.926}$
	SOS STO-3G	$N^{0.975}$
α_{zz}	FF STO-3G	$N^{1.663}$
	FF 4-31G	$N^{1.541}$
	SOS STO-3G	$N^{1.464}$
α	FF STO-3G	$N^{1.399}$
α	SOS STO-3G	$N^{1.285}$

This tabulation shows important and significant trends. First, the SOS values underestimate by 16–19% the corresponding values obtained by the FF methodology. This point will be discussed in detail in the second part of this chapter. The flexibility of the basis has a strong effect on the actual values of the components, as seen in Table I. Note that, as expected, the change is larger for the direction when the electrons are poorly described by minimal basis sets (e.g., α_{yy} corresponding to π electrons that are only described by $2p_yC$ orbitals in a minimal-basis-set calculation). A more extensive analysis on the ethylene molecule in the SOS scheme is given in Table II. This table includes results of minimal (STO-3G), extended (4-31G and 6-31G), and polarized (6-31G**) bases, indicating that the same trends are observed. Going from the minimal to the polarized basis yields an improvement by a factor of 1.86 in the total polarizability, the effect being more marked for the α_{yy}, which improves by a factor of 4.53. Let us note, however, that the "best" SOS calculated value (17.45 a.u.) is still only 61% of the "experimental" value (28.48 a.u.). This progressive improvement of the α value is analyzed in Fig. 3. The use of an extended basis has a strong effect on all the components of the polarizability. A better description of the 1sC inner shell has only a minor effect, which is more marked in the π direction. The inclusion of polarized orbitals still slightly affects the molecular in-plane components but markedly modifies the perpendicular α_{yy} component.

The above results, at both the FF and SOS levels, indicate that the components of the polarizability tensor follow qualitatively the same trends

TABLE I

Electric Polarizability Components[a]

	α_{xx}	α_{yy}	α_{zz}	α
Finite-field STO-3G				
C_2H_4	13.1964	2.5312	17.4773	11.0653
C_4H_6	24.8279	4.7491	47.8512	25.8096
C_6H_8	37.7190	6.9819	100.4982	48.3997
C_8H_{10}	49.5785	9.1377	174.5574	77.7579
Finite-field 4-31G				
C_2H_4	22.7813	5.9019	29.1906	19.2912
C_4H_6	40.7345	12.1874	73.3961	42.1060
C_6H_8	—	—	—	—
C_8H_{10}	—	—	—	—
SOS, STO-3G				
C_2H_4	11.1140	1.9293	15.1165	9.3866
C_4H_6	22.1237	3.7711	38.1210	21.3386
C_6H_8	33.8750	5.6385	73.7087	37.7387
C_8H_{10}	44.8826	7.4451	114.7559	55.6946
"Exp."				
C_2H_4				28.48[b]
C_4H_6				56.7[c]
C_6H_8				
C_8H_{10}				

[a] In a.u.: 1 a.u. $= 0.296352 \times 10^{-24}$ esu.
[b] Bridge and Buckingham (1966).
[c] Syrkin and Dyatkina (1964).

TABLE II

Effect of Basis Sets on the Polarizability of Ethylene as
Calculated with SOS

Component	STO-G	4-31G	6-31G	6-31G**
α_{xx} (a.u.)	11.1140	18.1468	18.4321	19.1702
α_{yy} (a.u.)	1.9293	5.6067	6.2169	8.7550
α_{zz} (a.u.)	15.1165	23.7830	24.3127	24.4237
α (a.u.)	9.3866	15.8455	16.3406	17.4496
	Exp.: 28.48a			

[a] Bridge and Buckingham (1966).

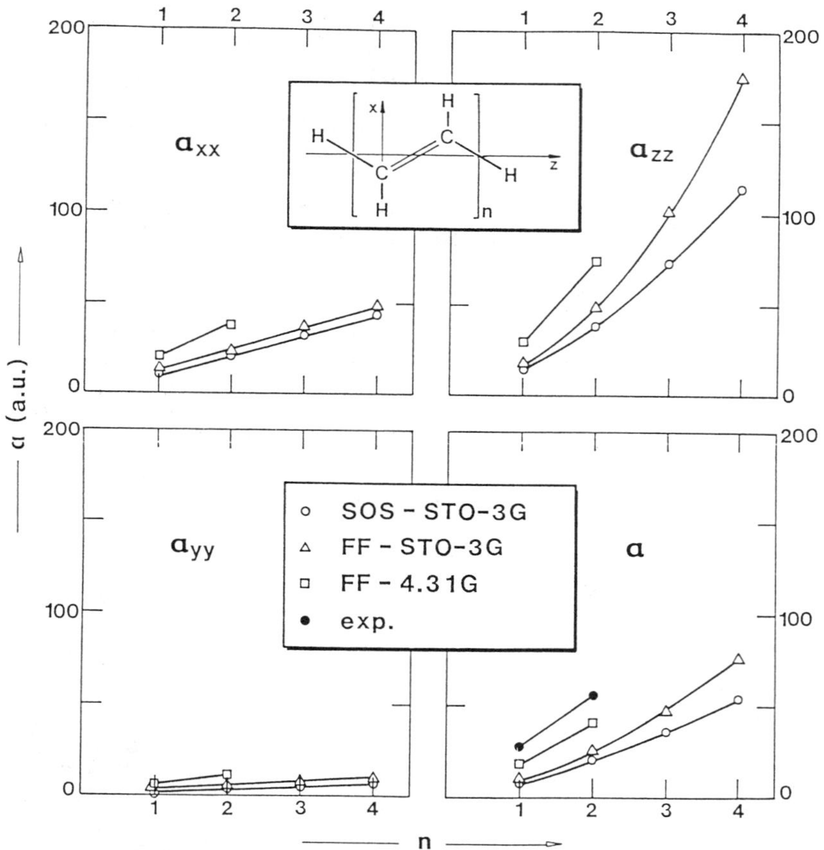

Fig. 2. Evolution of the three components and the average value of polarizability of polyene (C_nH_{n+2}) with respect to the chain length ($n = 1, 2, 3, 4$), calculated with the FF and SOS methods in two basis sets (STO-3G, 4-31G).

when calculated with the four basis sets. Therefore we use freely the STO-3G basis and the SOS method to investigate the influence of bond alternation on the longitudinal polarizability α_{zz} of different polyene chains. As in a previous study with the FF method (Bodart *et al.*, 1985b), calculations have been made for four alternation degrees while keeping the total chain length identical in order to properly separate alternation effects from other influences. The SOS results given in Table III are for the same compounds and geometries as in the above reference; the bond alternation is measured by the difference between the C—C and C=C bond lengths. Note that for all compounds, the values of the longitudinal polarizability are enhanced when the structure becomes more regular, and the effect is more pronounced in the larger compounds.

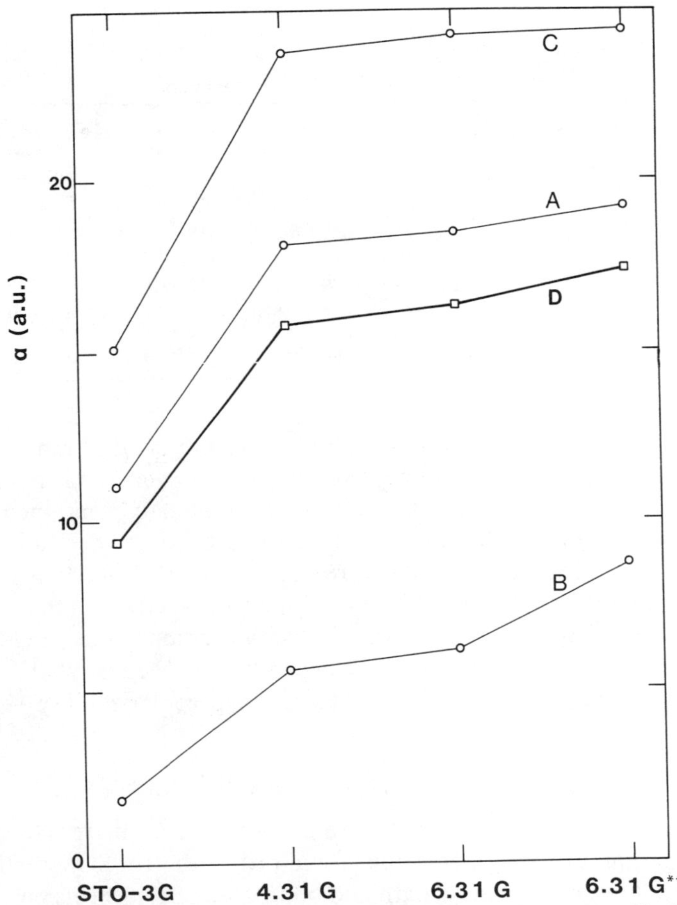

Fig. 3. Influence of the basis set on the polarizability of ethylene as calculated in SOS; $A:\alpha_{xx}$, $B:\alpha_{yy}$, $C:\alpha_{zz}$, $D:\alpha = \frac{1}{3}(\alpha_{xx} + \alpha_{yy} + \alpha_{zz})$.

TABLE III

Longitudinal Polarizability α_{zz} (a.u.) of Ethylene, Butadiene, Hexatriene, and Octatetraene in Four Different Geometries as Calculated with SOS

	Bond alternation $(R_{C-C} - R_{C=C})$ (Å)			
	0.1	0.05	0.025	0.0125
Ethylene	15.2708	15.9315	16.2802	16.4590
Butadiene	41.1084	45.0920	47.3616	48.5720
Hexatriene	75.2233	86.5056	93.3509	97.1279
Octatetraene	114.4393	137.1078	151.7125	160.0503

TABLE IV

Components and Average Value of the Second Hyperpolarizability (a.u.)
of Ethylene in Four Different Basis Sets Calculated in SOS

Component	STO-3G	4-31G	6-31G	6-31G**
γ_{zzzz}	-211.8563	-25.9414	-23.5789	58.4761
γ_{yyyy}	-2.7228	15.2426	27.1295	20.4483
γ_{xxxx}	4.0279	447.6782	459.2274	448.9096
γ_{zzyy}	43.9184	81.9738	94.5273	74.8330
γ_{yyxx}	22.0135	39.8581	46.9900	32.4285
γ_{zzxx}	-83.8498	-74.8124	-77.3086	-41.7629
γ	-49.2784	106.2037	118.2391	131.7663

A similar study of the basis-set, chain-length, and bond-alternation effects
on the components of the second hyperpolarizability tensor was not nearly as
complete at the time of writing this review and, as already mentioned, we were
not able to obtain reliable FF values for the third-order derivatives of the
dipole moment. The only available results are thus of the SOS type; they
are given in Tables IV and V. The first thing to observe from the content of
Table IV and Fig. 4 is how dramatically the various components of the
second hyperpolarizability tensor are dependent on the size of the basis set.
The value of the average (scalar part) second hyperpolarizability γ expressed
as

$$\gamma = \tfrac{1}{5}(\gamma_{xxxx} + \gamma_{yyyy} + \gamma_{zzzz} + 2\gamma_{xxyy} + 2\gamma_{yyzz} + 2\gamma_{xxzz}) \tag{22}$$

is therefore a complicated quantity to control and evaluate. Nevertheless,
assuming that the STO-3G results obtained with the SOS method can
meaningfully be used, chain-length and bond-alternation effects on γ_{zzzz} have
been tentatively assessed in the same way as for the longitudinal polarizability.

TABLE V

$zzzz$-Component Hyperpolarizability (a.u.) of Ethylene (E),
Butadiene (B), Hexatriene (H), and Octatetraene (O) in Four
Different Geometries

	Bond alternation $(R_{C-C}-R_{C=C})$ (Å)			
	0.1	0.05	0.025	0.0125
E	-252.4465	-312.2767	-345.7175	-363.3852
B	3280.4411	3606.9015	3702.7619	3723.5852
H	$23,114.529$	$30,156.655$	$33,972.032$	$35,868.851$
O	$75,329.345$	$112,197.66$	$135,705.51$	$148,607.11$

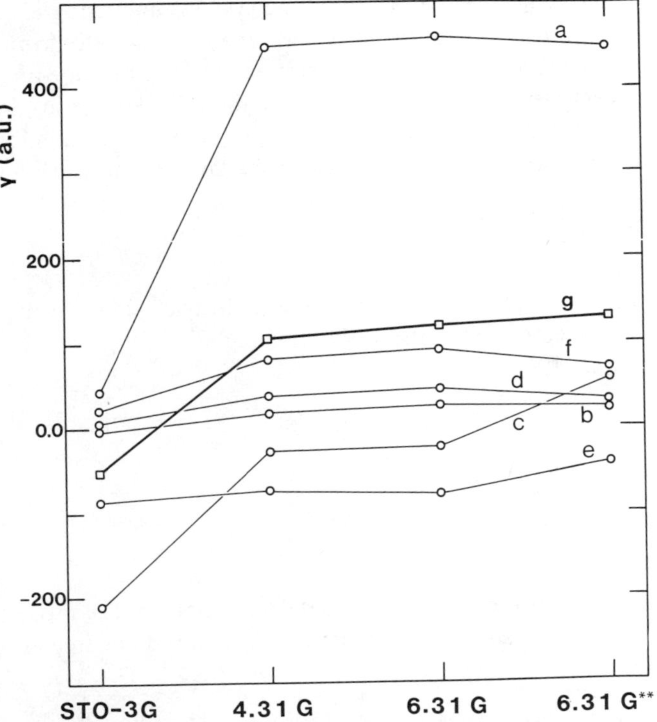

Fig. 4. Influence of the basis set on the hyperpolarizability of ethylene as calculated in SOS; a: γ_{xxxx}, b: γ_{yyyy}, c: γ_{zzzz}, d: γ_{xxyy}, e: γ_{xxzz}, f: γ_{yyzz}, g: $\gamma = \frac{1}{5}(\gamma_{xxxx} + \gamma_{yyyy} + \gamma_{zzzz} + 2\gamma_{xxyy} + 2\gamma_{yyzz} + 2\gamma_{xxzz})$.

Basically, the same trends as for α_{zz} are observed. The value of γ_{zzzz} increases very quickly with the size of the system, as was already known experimentally (Ward and Elliott, 1978) and theoretically (McIntyre and Hameka, 1978c; Papadopoulos *et al.*, 1982). Moreover, bond alternation plays an equally critical role as in the case of α_{zz}: the more regular the structure, the larger the value of γ_{zzzz}.

III. COMPARISON OF THE PHYSICAL CONTENT OF THE CHF AND SOS METHODS

The FF approach provides a practical way to perform coupled Hartree–Fock (CHF) calculations, while the SOS technique corresponds to the uncoupled Hartree–Fock (UCHF) method. As easily observed in Table I,

152   J.-M. André, C. Barbier, V. Bodart and J. Delhalle

both techniques when applied to the same systems and using identical basis sets do not yield the same results. This discrepancy originates from a different physical content. The purpose of this subsection is to provide an analysis of this difference. In the first part, the CHF and the UCHF versions are analyzed in detail, and in the second part we use the double-perturbation formalism to identify intermediate approaches between the UCHF and CHF levels.

A. CHF versus UHF

The perturbation associated with the static electric field being expressible as a sum of one-electron operators, the Hartree–Fock equations are written (Langhoff *et al.*, 1966) as

$$[h(1) + \lambda h'(1)]\phi_i(1) = \varepsilon_i \phi_i(1) \qquad i = 1, 2, \ldots, N \tag{23}$$

where

$$h'(1) = r \cdot F \tag{24}$$

$$\langle \phi_i | \phi_j \rangle = \delta_{ij} \tag{25}$$

Some care has to be exercised in writing the definition of $h(1)$. Two possibilities exist: the first one consists in inserting the unperturbed Hartree–Fock operator h^0 in Eq. (23),

$$h^0(1) = -\frac{1}{2}\nabla_1^2 + \sum_A \frac{Z_A}{r_{1A}} + \sum_{j=1}^{N} \left\langle \phi_j^0 \left| \frac{1 - P_{12}}{r_{12}} \right| \phi_j^0 \right\rangle \tag{26}$$

with

$$h^0(1)\phi_i^0(1) = \varepsilon_i^0 \phi_i^0(1) \tag{27}$$

The other possibility is to substitute, in h^0, the unperturbed spin orbitals ϕ_i^0 by the solutions ϕ_i of the perturbed Hartree–Fock expression, Eq. (23). Let us consider successively both cases.

1. First Case

The solutions of Eq. (23) are obtained by expanding the (unknown) spin orbitals ϕ_i and the orbital energies ε_i in a Taylor series in λ:

$$\phi_i = \phi_i^0 + \lambda \phi_i^1 + \lambda^2 \phi_i^2 + \cdots \tag{28}$$

$$\varepsilon_i = \varepsilon_i^0 + \lambda \varepsilon_i^1 + \lambda^2 \varepsilon_i^2 + \cdots \tag{29}$$

Upon introducing the relations of Eqs. (26), (28), and (29) in Eq. (23), we obtain

$$[h^0(1) + \lambda h'(1) - \varepsilon_i^0 - \lambda \varepsilon_i^1 - \cdots][\phi_i^0(1) + \lambda \phi_i^1(1) + \cdots] = 0 \tag{30}$$

which gives for the successive orders

$$[h^0(1) - \varepsilon_i^0]\phi_i^0(1) = 0 \qquad \text{(zero-order)} \qquad (31)$$

$$[h^0(1) - \varepsilon_i^0]\phi_i^1(1) + [h'(1) - \varepsilon_i^1]\phi_i^0(1) = 0 \qquad \text{(first-order)} \qquad (32)$$

and so on. By expanding the first-order spin orbitals ϕ_i^1 in terms of the solutions ϕ_i^0 of the unperturbed Hartree–Fock equations,

$$\phi_i^1(1) = \sum_p^{\text{unocc}} C_{pi}^1 \phi_p^0 \qquad (33)$$

we obtain the coefficients C_{pi}^1 and the second-order correction to the total energy ($E^{(2)}$) in their usual SOS formulation:

$$C_{pi}^1 = -\frac{\langle \phi_p^0 | h' | \phi_i^0 \rangle}{\varepsilon_p^0 - \varepsilon_i^0} \qquad (34)$$

$$E^{(2)} = -\sum_i \sum_p \frac{|\langle \phi_p^0 | h' | \phi_i^0 \rangle|^2}{\varepsilon_p^0 - \varepsilon_i^0} \qquad (35)$$

2. Second Case

As mentioned above, a more rigorous description, equivalent to the finite-field method in the limit of zero field, is obtained by using the perturbed spin orbitals ϕ_i to construct the operator $h(1)$ in Eq. (23). This operation introduces in the first-order relation, Eq. (32), two additional terms, which couple the first-order (ϕ_i^1) and the zero-order (ϕ_i^0) spin orbitals:

$$[h^0(1) - \varepsilon_i^0]\phi_i^1(1) + [h'(1) - \varepsilon_i^1]\phi_i^0(1)$$

$$+ \sum_{j=1}^N \left[\left\langle \phi_j^0 \left| \frac{1 - P_{12}}{r_{12}} \right| \phi_j^1 \right\rangle + \left\langle \phi_j^1 \left| \frac{1 - P_{12}}{r_{12}} \right| \phi_j^0 \right\rangle \right] \phi_i^0(1) = 0 \qquad (36)$$

This last equation, which has to be solved with respect to the normalization condition $\langle \phi_i | \phi_j \rangle = \delta_{ij}$, is known as the first-order coupled Hartree–Fock equation. When applying Eq. (33) into Eq. (36), the coefficients C_{pi}^1 are determined by the equation

$$C_{pi}^1(\varepsilon_p^0 - \varepsilon_i^0) + \langle \phi_p^0 | h' | \phi_i^0 \rangle$$

$$+ \sum_j \sum_q {}' C_{qj}^1 [\langle \phi_p^0 \phi_j^0 | \phi_i^0 \phi_q^0 \rangle - \langle \phi_p^0 \phi_j^0 | \phi_q^0 \phi_i^0 \rangle]$$

$$+ \sum_j \sum_q {}' C_{qj}^1 [\langle \phi_p^0 \phi_q^0 | \phi_i^0 \phi_j^0 \rangle - \langle \phi_p^0 \phi_q^0 | \phi_j^0 \phi_i^0 \rangle] = 0 \qquad (37)$$

To get the solutions of this equation—i.e., to compute the second-order correction energy—requires an iterative and time-consuming calculation because of the coupling terms.

As seen from the analysis, the CHF (or FF) approach is more rigorous than the UCHF technique in the sense that the average electron–electron interactions are treated self-consistently in the presence of the perturbation. This is corroborated by the results in Table I, where the FF values are consistently closer to the experimental values than are the SOS (UCHF) ones.

It might be interesting to indicate the existence of a scheme computationally less complex. It is obtained by the suppression in Eq. (36) of the coupling terms involving the first-order functions ϕ_j^1 other than the one being calculated (i.e., ϕ_i^1). This gives, as counterpart to Eq. (37),

$$C_{pi}^1(\varepsilon_p^0 - \varepsilon_i^0) + \langle \phi_p^0 | h' | \phi_i^0 \rangle + \sum_q C_{qi}^1 [\langle \phi_p^0 \phi_i^0 | \phi_i^0 \phi_q^0 \rangle - \langle \phi_p^0 \phi_i^0 | \phi_q^0 \phi_i^0 \rangle] = 0 \tag{38}$$

Moreover, if only the diagonal term ($p = q$) is kept in Eq. (38), the coefficients C_{pi}^1 are given by

$$C_{pi}^1 = \frac{\langle \phi_p^0 | h' | \phi_i^0 \rangle}{(\varepsilon_p^0 - \varepsilon_i^0) + \langle \phi_p^0 \phi_i^0 | \phi_i^0 \phi_p^0 \rangle - \langle \phi_p^0 \phi_i^0 | \phi_p^0 \phi_i^0 \rangle} \tag{39}$$

The denominator is then exactly the energy difference between the states corresponding to the excitation of an electron from spin orbital ϕ_i^0 to spin orbital ϕ_q^0. We have not been in a position to test this approximation on the polyene chains. However, according to calculations on atom polarizabilities (Langhoff $et\ al.$, 1966), results are intermediate between CHF and UCHF values, but require much less computing effort.

B. Double Perturbation Theory

Several authors (Caves and Karplus, 1969; Nakatsuji and Musher, 1974) analyzed the differences between the CHF and UCHF methods in terms of the double perturbation theory where the external field is one perturbation and the electron-correlation operator is the other. This second perturbation operator is defined as the difference between the true electron-repulsion potential (which takes place in the exact Hamiltonian) and the Hartree–Fock potential (which represents the average potential experienced by one electron in the field of the others):

$$V = \sum_j \frac{1}{r_{ij}} - \sum_j \left\langle \phi_j^0 \left| \frac{1 - P_{12}}{r_{12}} \right| \phi_j^0 \right\rangle \tag{40}$$

The aim of the double perturbation expansion is to express the CHF second-order energy in terms of successive electron-correlation corrections:

$$E_{\text{CHF}}^2 = E_0^2 + \lambda E_1^2 + \lambda^2 E_2^2 + \cdots \tag{41}$$

In this formulation, the leading term E_0^2 is nothing but the UCHF-SOS second-order (in field) energy [Eq. (35)]. We give in the appendix the detailed expression of the first-order correction E_1^2. One interesting feature of Eq. (41) is that it behaves roughly like a geometric series, which means that the exact value (i.e., E_{CHF}^2) can be approximated by an asymptotic expansion containing only the zero- and the first-order terms:

$$E_G^2 = E_0^2(1 - E_1^2/E_0^2)^{-1} \qquad (42)$$

Table VI shows how the values of the parallel and perpendicular polarizabilities of the hydrogen molecule are dependent on the different approximations of the correlation corrections. These results and the relative simplicity of the equations corresponding to the first-order correction, $E_0^2 + E_1^2$, or the geometric approximation, E_G^2, suggest that they could be applied to calculate reliable polarizabilities and hyperpolarizabilities of larger molecules.

IV. CONCLUDING REMARKS

According to the computational studies presented elsewhere (Bodart *et al.*, 1985b) and in this contribution, it is anticipated that the polarizability, normalized to the monomeric units, tends to reach an asymptotic limit, which grows when the systems exhibit increased geometrical regularity. For complex systems, this limit will soon be out of reach from studies on chains of increasing length. Thus, it would be very useful to be able to estimate this limit from calculations on infinite chains.

In going from molecules to polymers, we should at least mention the various perturbation-theoretic approaches that were developed for calculating the susceptibility tensors of solids (Butcher and McLean, 1963; Cheng and Miller, 1964; Genkin and Mednis, 1968). In particular, the linear (Cojan *et al.*, 1977) and nonlinear (Agrawal *et al.*, 1978) properties of one-dimensional conjugated polymers were studied using the Genkin–Mednis formalism, in the framework of the tight-binding approximation.

TABLE VI

Influence of the Correlation Corrections on the Parallel ($\alpha_{\parallel}$) and the Perpendicular ($\alpha_{\perp}$) Components of the Polarizability of the Hydrogen Molecule[a]

Component	E_0^2 (UCHF)	$E_0^2 + E_1^2$	E_G^2	E^2 (CHF)	Exp.
$\alpha_{\parallel}$	0.721	0.896	0.949	0.960	1.03
$\alpha_{\perp}$	0.459	0.624	0.678	0.682	0.72

[a] In cubic angstroms: $1 \text{ Å}^3 = 10^{-24}$ esu.

Since these methods have already been reviewed elsewhere (Flytzanis, 1975), the present section briefly points to some less-known results concerning the applicability of the finite field to large periodic systems. One could think this as being rather trivial, merely replacing field-dependent molecular orbitals by field-dependent Bloch crystalline orbitals. However, as shown by Churchill and Holmstrom (1982, 1983), serious difficulties arise in imposing realistic boundary conditions to solve the one-electron eigenvalue equation; under the boundary conditions commonly used in treating the zero-field case (e.g., Born–von Karman boundary conditions), this equation either leads to physically inconsistent results or, still worse, has no solution at all!

The most striking fact that emerges from these studies is that the choice of boundary conditions, which has quite little influence in the treatment of unperturbed large periodic systems, is of paramount importance in considering the non-zero-field case. The boundary conditions traditionally used in solid-state physics turn out to lead to spectacular inconsistencies. Even with realistic—e.g., box—boundary conditions, the weak-field and long-chain limits cannot be taken in arbitrary order. This unusual behavior is a consequence of the pathological nature of the perturbing term, eFr, which becomes undeterminate in the limits as $F \to 0$ and $r \to +\infty$. This does not definitively rule out infinite lattices, however, since the perturbation induced by a uniform dc field can always be considered as the long-wavelength limit of a spatially periodic, time-independent perturbation (Kunc and Resta, 1983).

ACKNOWLEDGMENTS

The authors are indebted to Dr. J. Zyss and Dr. J. G. Fripiat for fruitful discussions. They acknowledge with appreciation the support of their collaboration under the ESPRIT-EEC contract no. 443 on molecular engineering for optoelectronics and under NATO grant no. 130.83. One of us (V.P.B.) is grateful to the Institut pour l'Encouragement de la Recherche Scientifique dans l'Industrie et l'Agriculture (IRSIA, Belgium) for financial support.

APPENDIX

We report here the deduction of the first-order correlation correction to the second-order (in electric field) energy, E_1^2, given in Eq. (41). The starting point of this calculation is the classical third-order correction to the total energy E^3, which, in terms of many-electron states, is written as

$$E^3 = \sum_K{}' \sum_L{}' \frac{\langle 0|W|K\rangle\langle K|W|L\rangle\langle L|W|0\rangle}{(E_0 - E_K)(E_0 - E_L)} - \langle 0|W|0\rangle \sum_K{}' \frac{|\langle 0|W|K\rangle|^2}{(E_0 - E_K)^2} \quad (43)$$

where W is the perturbation operator and the prime on the summation means that the terms $n = 0$ and $m = 0$ are excluded. In the double-perturbation theory, the operator W is a sum of two different operators,

$$W = h' + V \tag{44}$$

where h' is the dipolar operator and V is the correlation operator [see Eq. (40)]. Substituting Eq. (44) in Eq. (43), we obtain an expression for E^3 that can be divided in four terms,

$$E^3 = C^3 + FC^2 + F^2C + F^3 \tag{45}$$

where

$$C^3 = \sideset{}{'}\sum_K \sideset{}{'}\sum_L \frac{\langle 0|V|K\rangle\langle K|V|L\rangle\langle L|V|0\rangle}{(E_0 - E_K)(E_0 - E_L)} - \langle 0|V|0\rangle \sideset{}{'}\sum \frac{|\langle 0|V|K\rangle|^2}{(E_0 - E_K)^2} \tag{46}$$

$$FC^2 = \sideset{}{'}\sum_K \sideset{}{'}\sum_L [\langle 0|V|K\rangle\langle K|V|L\rangle\langle L|h'|0\rangle + \langle 0|V|K\rangle\langle K|h'|L\rangle\langle L|V|0\rangle$$

$$+ \langle 0|h'|K\rangle\langle K|V|L\rangle\langle L|V|0\rangle] \; \frac{1}{(E_0 - E_K)(E_0 - E_L)}$$

$$- \langle 0|h'|0\rangle \sideset{}{'}\sum_K \frac{|\langle 0|V|K\rangle|^2}{(E_0 - E_K)^2} - 2\langle 0|V|0\rangle \sideset{}{'}\sum_K \frac{\langle 0|V|K\rangle\langle 0|h'|K\rangle}{(E_0 - E_K)^2} \tag{47}$$

$$F^2C = \sideset{}{'}\sum_K \sideset{}{'}\sum_L [\langle 0|h'|K\rangle\langle K|h'|L\rangle\langle L|V|0\rangle + \langle 0|h'|K\rangle\langle K|V|L\rangle\langle L|h'|0\rangle$$

$$+ \langle 0|V|K\rangle\langle K|h'|L\rangle\langle L|h'|0\rangle]\frac{1}{(E_0 - E_K)(E_0 - E_L)}$$

$$- \langle 0|V|0\rangle \sideset{}{'}\sum_K \frac{|\langle 0|h'|K\rangle|^2}{(E_0 - E_K)^2} - 2\langle 0|h'|0\rangle \sideset{}{'}\sum_K \frac{\langle 0|h'|K\rangle\langle 0|V|K\rangle}{(E_0 - E_K)^2} \tag{48}$$

$$F^3 = \sideset{}{'}\sum_K \sideset{}{'}\sum_L \frac{\langle 0|h'|K\rangle\langle K|h'|L\rangle\langle L|h'|0\rangle}{(E_0 - E_K)(E_0 - E_L)} - \langle 0|h'|0\rangle \sideset{}{'}\sum \frac{|\langle 0|h'|K\rangle|^2}{(E_0 - E_K)^2} \tag{49}$$

In this notation, the term F^iC^j corresponds to the correction at the ith order in the electric field (operator h') and to the jth-order correlation correction. Accordingly, E_1^2 is equal to the term F^2C^1, that is,

$$E_1^2 = \sideset{}{'}\sum_K \sideset{}{'}\sum_L [\langle 0|h'|K\rangle\langle K|h'|L\rangle\langle L|V|0\rangle + \langle 0|h'|K\rangle\langle K|V|L\rangle\langle L|h'|0\rangle$$

$$+ \langle 0|V|K\rangle\langle K|h'|L\rangle\langle L|h'|0\rangle]\frac{1}{(E_0 - E_K)(E_0 - E_L)}$$

$$- \langle 0|V|0\rangle \sideset{}{'}\sum_K \frac{|\langle 0|h'|K\rangle|^2}{(E_0 - E_K)^2} - 2\langle 0|h'|0\rangle \sideset{}{'}\sum_K \frac{\langle 0|h'|K\rangle\langle 0|V|K\rangle}{(E_0 - E_K)^2} \tag{50}$$

REFERENCES

Agrawal, G. P., Cojan, C., and Flytzanis, C. (1978). *Phys. Rev. B* **17**, 776.
Atkins, P. W. (1983). "Molecular Quantum Mechanics." Oxford Univ. Press, London and New York.
Bloembergen, N. (1965). "Nonlinear Optics." Benjamin, New York.
Bodart V. P., Delhalle, J., André, J. M., and Zyss, J. (1985a). *Can. J. Chem.* **63**, 1631.
Bodart V. P., Delhalle, J., André, J. M., and Zyss, J. (1985b). *In* "Polydiacetylenes: Synthesis, Structure and Electronic Properties" (D. Bloor and R. R. Chance, eds.). Martinus Nijhof, Dordrecht, Netherlands, pp. 125–133.
Bridge, N. J., and Buckingham, A. D. (1966). *Proc. R. Soc. London, Ser. A* **295**, 334.
Butcher, P. N. (1965). "Nonlinear Optical Phenomena." Ohio State University, Columbus.
Butcher, P. N., and McLean, T. P. (1963). *Proc. Phys. Soc. London* **81**, 219.
Caves, T. C., and Karplus M. (1969). *J. Chem. Phys.* **50**, 3649.
Cheng, H., and Miller, P. B. (1964). *Phys. Rev.* **134**, A683.
Churchill, J. N., and Holmstrom, F. E. (1982). *Am. J. Phys.* **50**, 848.
Churchill, J. N., and Holmstrom, F. E. (1983). *Physica B (Amsterdam)* **123**, 1.
Cohen, H. D., and Roothaan, C. C. J. (1965). *J. Chem. Phys.* **43**, 534.
Cojan, C., Agrawal, G. P., and Flytzanis, C. (1977). *Phys. Rev. B* **15**, 909.
Davies, P. L. (1952). *Trans. Faraday Soc.* **47**, 789.
Delhalle, J., Bodart, V. P., Dory, M., André J. M., and Zyss, J. (1985). *Int. J. Quantum Chem. Symp.* **19**, 313.
Ditchfiel, R., Hehre, W. J., and Pople, J. A. (1971). *J. Chem. Phys.* **54**, 724.
Flytzanis, C. (1975). *In* "Quantum Electronics. A Treatise" (H. Rabin and C. L. Tangs, eds.), Vol. 1A, Part A, Chapter 2. Academic Press, New York.
Genkin, V. N., and Mednis, P. M. (1968). *Sov. Phys.—JETP (Engl. Transl.)* **27**, 609.
Hameka, H. F. (1977). *J. Chem. Phys.* **67**, 2935.
Hariharan, P. C., and Pople, J. A. (1973). *Theor. Chim. Acta* **28**, 213.
Hehre, W. J., Stewart, R. F., and Pople, J. A. (1969). *J. Chem. Phys.* **51**, 2657.
Hermann, J. P., and Ducuing, J. (1974). *J. Appl. Phys.* **45**, 5100.
Kunc, K, and Resta, R. (1983). *Phys. Rev. Lett.* **51**, 686.
Langhoff, P. W., Karplus, M., and Hurst, R. P. (1966). *J. Chem. Phys.* **44**, 505.
McIntyre, E. F., and Hameka, H. F. (1978a). *J. Chem. Phys.* **68**, 3481.
McIntyre, E. F., and Hameka, H. F. (1978b). *J. Chem. Phys.* **68**, 5534.
McIntyre, E. F., and Hameka, H. F. (1978c). *J. Chem. Phys.* **69**, 4814.
McIntyre, E. F., and Hameka, H. F. (1979). *J. Chem. Phys.* **70**, 2215.
Nakatsuji, H., and Musher, J. I. (1974). *J. Chem. Phys.* **61**, 3737.
Papadopoulos, M. G., Waite, J., and Nicolaides, C. A. (1982). *J. Chem. Phys.* **77**, 2527.
Pople, J. A., and Hehre, W. J. (1978). *J. Comput. Phys.* **27**, 161.
Roothan, C. C. J. (1951). *Rev. Mod. Phys.* **23**, 69.
Rustagi, K. C., and Ducuing, J. (1974). *Opt. Commun.* **10**, 258.
Rutishauser, H. (1963). *Numer. Math.* **5**, 48.
Syrkin, Y. K., and Dyatkina, M. E. (1964). "Structure of Molecules and Chemical Bond," p. 201. Dover, New York.
Szabo, A., and Ostlund, N. S. (1982). "Modern Quantum Chemistry: Introduction to Advanced Electronic Structure Theory." Macmillan, New York.
Ward, J. F., and Elliott, D. S. (1978). *J. Chem. Phys.* **69**, 5438.
Zamani-Khamiri, O., and Hameka, H. F. (1979). *J. Chem. Phys.* **71**, 1697.
Zamani-Khamiri, O., and Hameka, H. F. (1980). *J. Chem. Phys.* **73**, 1980.
Zamani-Khamiri, O., McIntyre, E. F., and Hameka, H. F. (1980a). *J. Chem. Phys.* **72**, 1280.
Zamani-Khamiri, O., McIntyre, E. F., and Hameka, H. F. (1980b). *J. Chem. Phys.* **72**, 5906.

Chapter III-6

Resonant Molecular Optics

B. DICK

Max Planck Institut für biophysikalische Chemie
Karl-Friedrich-Bonhoeffer Institut, Abteilung Laserphysik
Postfach 2841, D-3400 Göttingen, Federal Republic of Germany

R. M. HOCHSTRASSER

Department of Chemistry
University of Pennsylvania
Philadelphia, Pennsylvania 19104

H. P. TROMMSDORFF

Laboratoire de Spectrométrie Physique associé au C.N.R.S.;
Université Scientifique, Technologique, et Médicale de Grenoble
BP 87, 38402 St. Martin d'Hères Cedex, France

I. INTRODUCTION

Molecular optics concerns the study of nonlinear processes and materials derived from molecular systems. The existence of a wide range of molecular structures gives rise to a correspondingly large variety of optical transitions. These transitions, or resonances, determine the nature of the responses of the

159

system to electromagnetic fields; hence their detailed study is of paramount importance to the development of a fuller understanding of the nonlinear properties. For some purposes it is advantageous to optimize the nonlinear response in the transparent regime of the material. In such a case the losses are minimal and the response times are shortest. However, enormous enhancements of the nonlinear signals occur when the incident electromagnetic field frequencies match those of the optical transitions. In order to understand the responses in such cases, it is necessary to have detailed knowledge of both the optical transitions and the dynamical processes that can be undergone by the excited states. This special case of *resonant molecular optics* is the subject of the present chapter.

In order to lay the basis for later applications to specific systems, Section II consists of a survey of the structural and dynamical properties of molecular condensed matter. Both the vibrational and the electronic resonances of molecular solids are discussed, as are the effects on them of intermolecular interactions, energy transfer, charge transfer, and disorder. Section III contains the theoretical foundation of the resonant nonlinear responses, from which later results can be deduced. Density matrix methods that can be used to describe the interaction of molecules with both weak and strong laser fields are developed in this section, which relies heavily on previous fundamental work on the resonant nonlinear response (Bloembergen, 1965; Butcher, 1965; Hellwarth, 1977; Flytzanis, 1975). In resonant nonlinear experiments, energy is transferred from the electromagnetic field to the medium, so in Section III we also consider the optics of each experiment in terms of complex wave vectors including all the relevant absorption coefficients. Section IV is the main part of the chapter, consisting of a review of resonant nonlinear optical experiments chosen to illustrate the variety of molecular responses and systems that have been studied during the past few years. The nonlinear optical properties and spectroscopic transitions can usually be studied by the time-domain or frequency-domain methods. A part of our interest has been to establish experimental and theoretical relationships between these two approaches, and such comparisons are brought out whenever possible.

II. STRUCTURAL AND DYNAMICAL PROPERTIES OF MOLECULAR SYSTEMS

A. Electronic States and Transitions

The linear absorption strengths of electronic transitions in molecules range over ~ 12 orders of magnitude. The weakest are the singlet-to-triplet transitions, which may have f values as low as 10^{-12}, and the strongest spin-allowed transitions have $f \approx 1$. The range of nonlinear interaction strengths is even greater—10^{24} for second-order and 10^{36} for third-order processes. Even the symmetry-allowed transitions of a molecule cover a range of f values that for an aromatic hydrocarbon might be as much as two orders of magnitude. Thus it is apparent that the extent of resonant enhancement of optical processes is very much dependent on the specific nature of the resonant state; for example, it is quite possible to have an exact resonance with a weak transition and yet have the signal dominated by nonresonant background effects. However, the extent of the chemical and structural changes that occur on electronic excitation is not necessarily related to the transition strength. For example, large changes in permanent dipole moment are known to occur for both weak and strong transitions.

The most common types of molecular excited states involve two-electron configurations of the type $\pi\pi^*$, $\sigma\pi^*$, and $n\pi^*$, or the excited electron is moved to another region of the molecule, leaving behind a positive center, to form a charge-transfer state. The low-energy states of aromatic and polyene molecules are of $\pi\pi^*$ type, whereas molecules having valence-shell electrons that are nonbonding frequently have low-energy $\sigma\pi^*$, $n\pi^*$, or charge-transfer states. Aromatic molecules and polyenes are often approximately centrosymmetric, such that their states can be classified as even (g) and odd (u) functions of the electronic coordinates. The aromatics usually have the u states at lowest energy, but there are many exceptions to this, including biphenyl (Whiteman *et al.*, 1973) and biphenylene (Hochstrasser and McAlpine, 1966; McAlpine, 1968), both of which have low-energy g states. Polyenes longer than three double bonds generally have a g state at lowest energy (Hudson *et al.*, 1982). Since these symmetry designations apply only to the equilibrium nuclear configurations of the molecules and there are many degrees of freedom, it follows that no electronic transitions are strictly forbidden: the zero-point motion can always allow the nuclei to seek out a configuration that is sufficiently distorted for the transition to become allowed. The magnitude of this so-called Herzberg–Teller contribution to the absorption coefficient depends on a variety of factors, including the proximity of intense electronic transitions for molecules having the equilibrium structure. Typically, for aromatics the extinction coefficient $\varepsilon_{\max}$ of such induced spectra is ~ 500 liters mol^{-1} cm^{-1} (absorption cross section σ of $\sim 2 \times 10^{-18}$ cm^2), whereas the allowed parts, when they occur, have $\varepsilon_{\max} \approx 10^4$ liters mol^{-1} cm^{-1} ($\sigma \approx 4 \times 10^{-17}$ cm^2). While the allowed $\pi \to \pi^*$ transitions of aromatics and polyenes are relatively strong, the symmetry-allowed $n \to \pi^*$ transitions, because of the small overlap of the electronic orbitals involved, are often 10–100 times less intense. Also intense are some transitions where the charge is substantially displaced in the excited state. These include many polar dyes, where the optical transition approximately corresponds to a transformation into a quinoidal structure, and conjugated systems having separated electron donor (such as dimethylamine) and acceptor (such as nitro) substituents. Another example of very strong transitions occurs for weak charge-transfer complexes (Mulliken charge transfer), where an electron is transferred completely from the donor to the acceptor on excitation (Mulliken and Person, 1975). While such materials are not as useful in applications where transparency is required, they are excellent candidates for the exploration of resonant nonlinear processes.

The difference between the ground- and excited-state dipole moments $\Delta\mu$ in these cases that involve substantial electron transfer can be extremely large— 20 D is not uncommon (Liptay, 1974)—so that this contribution to the magnitude of the nonlinear response can be dominant. The excited-state dipole moments are readily measured in solids by studies of spectra in electric fields (Hochstrasser, 1973). Moderately large values of $\Delta\mu$ of ~ 3 D are found for $n \to \pi^*$ transitions in heteroaromatics and ketones, but $\Delta\mu \approx 0$ for aromatics even if they are in principle polar.

Besides transitions from the ground states, the spectra of excited organic molecules have also been studied. These transitions are very important to the nonlinear response, since they correspond to the intermediate steps, which occur as either real or virtual transitions, in the multiphoton processes. However, there does not exist such a large data set for excited-state transitions, and there is still a strong reliance on theoretical calculations to predict this important information. Experiments have exposed a few empirical rules that are useful. The excited-state absorption spectra of aromatic molecules resemble the spectra of the corresponding negative ions that have strong absorption in the visible and near ultraviolet. The $n\pi^*$ states have strong absorption bands at low energy, corresponding to changes in the π-electron distribution. Similarly, charge-transfer states have strong transitions to higher-energy neutral states and excited states of the negative and positive ion components. Members of still another class of molecules, the polyphenyls, have excited states with dye-like absorption spectra as a result of the greatly increased conjugation and tendency toward a planar configuration in the excited state compared with the normally nonplanar ground states.

B. Crystal Spectra—Excitons

The electronic and vibrational spectra of molecules are changed on passing to the solid phase. Molecular solids normally contain two or more molecules in each unit cell, so that the crystals consist of two or more interpenetrating lattices. Most neutral molecules form either van der Waals lattices or in addition may form networks through weak chemical interactions such as hydrogen bonds. Since there is very little electron exchange between the molecules in the structure, the electronic properties of most molecular crystals are adequately described as slightly perturbed versions of the molecular properties. There are of course exceptions to this, including the charge-transfer salts, but this chapter will mainly refer to van der Waals or hydrogen-bonded solids.

The molecular excitations can transfer within the crystal, giving rise to excitation bands (Frenkel excitons) of width β. The excitation transfer time is on the order of $\hbar/\beta$, and for many of the most common crystals this time is long compared with the internal vibration periods of the molecules. In this case the Frenkel excitons correspond to the excitations of a slightly perturbed array of approximately harmonic oscillators. Each molecular electronic–vibrational excitation then has its own distinct exciton band. For the ground electronic state, the excitations are usually termed vibrons and β is typically 1–20 cm^{-1}, dependent on whether the vibrational motion corresponds to an oscillating polarizability or an oscillating dipole, the former being Raman-active and the latter for infrared-active (IR-active) modes. The IR-active modes show the largest vibron bandwidths. For electronically excited states, the exciton bands are formed by the electronic interaction, and the transfer of the vibrational energy alone represents a minor perturbation. The bandwidths are then determined by apportioning the electronic interaction in accordance with the Franck–Condon factor of the transition from the ground state to the level in question. In certain cases of very strong optical transitions, the electronic interaction is sufficiently large that new surfaces for nuclear motion are formed in the crystal! That is the so-called strong coupling case, for which it is no longer appropriate to consider the vibrational motion as slow compared with energy transfer. Only weak coupling occurs for molecular ground states and for the excited states of most simple molecular solids consisting of aromatics and their substituted derivatives. The weak coupling also applies when the excited-state equilibrium geometry differs significantly from the ground state, such as for weak charge-transfer complexes.

The energies of transitions to molecular excited states in crystals (Davydov, 1971) for the weak coupling case are given by

$$\Delta E = \Delta\varepsilon + D + \beta(k) \tag{1}$$

where $\Delta\varepsilon$ is the gas-phase transition energy and D is the gas-to-crystal shift resulting from intermolecular interactions that do not involve exchanging the relevant excitation. The exciton band term $\beta(k)$ describes the energy as a function of the wave vector for the relevant excitation. For nearest-neighbor interactions (Robinson, 1970) and a so-called restricted Frenkel limit in which certain small terms are excluded, $\beta(k)$ for a crystal having two molecules per unit cell has the form

$$\beta(k) = \sum_a 2\beta_{aa}\cos(k_a a) \pm \sum_{a<b} 4\beta_{ab}\cos\left(\frac{k_a a}{2}\right)\cos\left(\frac{k_b b}{2}\right) \tag{2}$$

where β_{ij} (with $i,j = a,b,c$) is the matrix element for transfer between the two sublattices and β_{aa} is the nearest-neighbor transfer along the a chain, and so on. The optical selection rule is $\Delta k \approx 0$ so that spectra show a splitting of the molecular transitions (Davydov splitting), which depends on the number of molecules in the unit cell. These $k \approx 0$ states form a basis for irreducible representations of the factor group of the equilibrium lattice space group, so that symmetry designations are extremely useful in describing the spectroscopic transitions and selection rules.

C. Impurity Spectra

Both chemical and isotopic impurities are readily dissolved in molecular solids to form substitutional solid solutions. Such impurities alter the optical properties through a variety of mechanisms. Most importantly, the band structures are perturbed, resulting in qualitative changes in the exciton dynamics. The optical spectra of the impurities resemble those of the free molecules. For example, the isotopic impurities have transition energies

$$\Delta E = \Delta\varepsilon + D \tag{3}$$

The selection rules for the impurity spectra are determined by the symmetry of the crystal site at which the molecule is located. It is usual that the lineshapes of impurity optical transitions contain both homogeneous and inhomogeneous contributions. The inhomogeneous contributions are most evident at low temperatures, where the limiting widths of optical transitions are in the range of a few reciprocal centimeters. Hole-burning and photon-echo experiments have proven the character of these low-temperature lines (de Vries and Wiersma, 1976; Small, 1983; Hesselink and Wiersma, 1983) but little was yet learned about the origin or specific properties of the inhomogeneties. The homogeneous contributions to the linewidths involve both pure dephasing and population relaxation contributions, which are quite system-dependent. At low temperatures the pure dephasing is induced by local lattice excitations causing fluctuations at the impurity center, whereas the T_1 relaxations are often approximately independent of temperature.

D. Inhomogeneous Broadening—Line Shapes

The exact nature of the strain disorder in molecular crystals is not understood. It has been suggested that the strain field arises from extended dislocations produced during the freezing process. The spectrum of such disorder must be characterized by both a strength and a correlation length. Simple theoretical models have been proposed to account for dephasing due to this kind of disorder. Klafter and Jortner (1977, 1978) derived an approximate lineshape theory for the absorption spectrum of triplet excitons (Hochstrasser, 1976). Abram and Hochstrasser (1980) have treated directly the time evolution of vibron coherence. Both of these theories use a simple tight-binding Hamiltonian to which a site diagonal perturbation is added to describe the effect of the crystal strain field on the site energies. The pure crystal exciton (vibron) band is characterized by a width W. The site energy perturbations are taken to be Gaussian random variables with a correlation length equal to the lattice spacing, and mean squared value σ:

$$\langle \Delta_i \Delta_j \rangle = \delta_{ij} \sigma^2 \tag{4}$$

These theories predict that

(1)   When $\sigma > W$, the band is inhomogeneously broadened.

(2)   When $\sigma \ll W$, the linewidth is much narrower than σ.

(3)   When $\sigma \ll W$ and the $k = 0$ state lies near a band singularity, the lineshape is asymmetric and the corresponding coherence decay is strictly nonexponential.

The magnitude of σ can be evaluated from the lineshape of a dilute isotopic impurity in the host material, assuming that the lineshape is not dominated by lifetime broadening, and that the impurity level is sufficiently separated from the host bands. This point is related to point (1) above in that dilute impurities form a (random) sublattice with vanishing exciton interaction, hence, $W \approx 0$. Point (2) was first layed out theoretically by the calculations of Klafter and Jortner (1978). An explicit discussion of this "motional narrowing" effect was given by Abram and Hochstrasser (1979), who pointed out that the overall coherence loss could be viewed as a competition between the site diagonal disorder and the correlation forced on the site amplitudes by intermolecular (excitonic) coupling. The term σ is the upper bound to the linewidth in the weak disorder model, and the introduction of exchange coupling makes the linewidth smaller. It was then suggested (DeCola *et al.*, 1980b) that the existence of motional narrowing in molecular crystals makes it possible for lifetime broadening to dominate the linewidth of the vibrational transitions even when the intrinsic disorder width is much larger than the population decay rate. The T_1 processes are also predicted to be influenced by the disorder (Velsko and Hochstrasser, 1985a,b).

E. Relaxation Processes

The nuclear dynamics of molecular solids can be conceptualized by considering the internal and external motions to be separate. With the exception of very-low-frequency vibrations, the internal modes of aromatic crystals might contain very little external motional character. The relaxation processes can then be understood, in a first approximation, by weak coupling of harmonic oscillators. The expectation exists, and this was confirmed theoretically in a few cases (Righini *et al.*, 1983), that the relaxation pathways can be understood by considering the low-order derivatives of the inter-molecular potential. Processes involving one lattice mode are expected to dominate. The lattice frequencies extend to ~ 200 cm^{-1} so that these one-phonon anharmonic effects should be dominated by the third-order mixed mode term $\gamma_3 = (\partial^3 V/\partial Q_1, \partial Q_2\, \partial q)_0$ where Q_1 and Q_2 are internal modes and q involves the relative motion of molecules. The relaxation time τ is then given by a Golden rule for each energetically allowed pathway. It is evident that the relaxation of ground-state vibrational levels then corresponds to motion in a set of coupled exciton bands, with the band from one internal mode acting as the trap for vibrational energy in other modes. The dynamics of vibrations in excited electronic states is more complex, since in addition the vibrational and electronic energy can separate with the difference being absorbed or emitted by the external phonons, depending on the temperature (Hochstrasser *et al.*, 1979).

III. BASIC THEORY

A. Introduction

The various ways that molecular systems can interact with optical radiation fields can be classified into the two main categories of dissipative and parametric processes. Dissipative processes exchange energy between the molecules and the light field through absorption and emission. In parametric processes the quantum state of the system is not changed, but energy and momentum are exchanged between different modes of the light field. This conservation of energy and momentum within the field components implies that the interaction of light beams having well-defined frequencies and directions will lead to new beams with new frequencies travelling again in well-defined directions. To use a chemical analogy, in parametric processes the molecules assume the role of catalysts.

There are two limiting methods by which nonlinear optical experiments are usually performed. In the frequency-domain approach, the response of the molecular system under investigation is measured as a function of the frequencies of the applied laser fields. The result is a spectrum consisting of resonance lines with certain center frequencies, amplitudes, and widths. In the alternative time-domain methods, the frequencies of the lasers are fixed, and the time delay between the very short laser pulses having wide-frequency bandwidths is varied. The resulting decay curves measure directly the relaxation processes in the molecule. For frequency-domain experiments, monochromatic waves provide information about all the system relaxations, whereas for time-domain experiments the ideal light pulses must impact on the system faster than all relaxation processes for the same goal to be achieved. Real laser sources have finite widths both in frequency and in time. In resonant nonlinear processes—which all spectroscopic arrangements are by definition—the choice of "ideal sources" for frequency-domain experiments depends on the complexities of the damping of the coupled levels. The choice of laser pulse width in relation to the damping rates selects a particular timescale, which determines the effective susceptibility of the system. Obviously, slow processes do not contribute to changes occuring during short light pulses.

B. Nonlinear Polarization, Response Function, and Susceptibility

The key quantity for the understanding of both parametric and dissipative processes is the polarization $\mathbf{P}(\mathbf{r}, t)$ induced in the sample by the electric field $\mathbf{E}(\mathbf{r}, t)$ of the light beams. Although nonlinear optical experiments require laser beams with high powers, the interaction energy is often still small enough compared with intramolecular energies to allow a treatment of the problem with perturbation theory even when the molecules are excited on resonance (see, however, Section III.H). This in turn allows a Taylor-series expansion of the induced polarization:

$$\mathbf{P}(\mathbf{r}, t) = \mathbf{P}^{(1)}(\mathbf{r}, t) + \mathbf{P}^{(2)}(\mathbf{r}, t) + \mathbf{P}^{(3)}(\mathbf{r}, t) + \cdots \tag{5}$$

Here we are concerned with the second and third term of this expansion. The first term describes linear absorption, light emission, refraction, and reflection. The second term refers to coherent sum and difference frequency generation. There are no dissipative processes associated with even-order nonlinear polarizations, except when one of the applied fields is a dc field. The third term is the source of two-photon absorption, Raman scattering, and a variety of coherent spectroscopies like CARS, CSRS, coherent Rayleigh scattering, polarization spectroscopy, phase conjugation, and most pump-probe spectroscopic methods.

In molecules and molecular crystals, the interaction with the laser field can be regarded as local in space, although nonlocal in time. This means that the polarization $\mathbf{P}(\mathbf{r}, t)$ induced at a certain point in space and time depends on the electric field strength $\mathbf{E}(\mathbf{r}, t')$ at that same point in space at all times $t' \leq$ time t. In a general way this relation is expressed as a convolution with a molecular response function $R^{(n)}$ (Butcher, 1965):

$$P_i^{(n)}(t) = \int_{-\infty}^{+} \int_{-\infty}^{+} \cdots \int_{-\infty}^{+} dt_1\, dt_2 \cdots dt_n\, R_{ijk\ldots l}^{(n)}(t - t_1, t - t_2, \ldots, t - t_n$$

$$\times\, E_j(t_1)E_k(t_2)\cdots E_l(t_n) \tag{6}$$

The response function $R^{(n)}$ is a tensor of rank $n + 1$, and the indices $ijkl$ refer to cartesian coordinates, with summation implied for doubly occuring indices. A physically meaningful response function must be real and vanish for any time argument approaching the remote past. This latter fact means that the molecules have a finite memory. The term $\mathbf{E}(t)$ is the total electric field interacting with the sample and can consist of a series of pulses centered at different times. Then the product EEE in $P^{(3)}$ contains several expressions each related to a particular pulse sequence of up to three pulses.

The ideal light pulses for time-domain spectroscopy have pulse envelopes much shorter than the characteristic time constants of the response function. If we model them by delta functions,

$$\mathbf{E}(t) = \sum_n \mathbf{E}_n = \sum_n \mathscr{E}_n e^{i\omega_n t}\, \delta(t - \tau_n) \tag{7}$$

then the contribution to the polarization $\mathbf{P}^{(3)}$ resulting from the δ-pulse sequence $E_1 E_2^* E_3$ $(\tau_1 < \tau_2 < \tau_3)$ is

$$\mathbf{P}^{(3)}(t) = R^{(3)}(t - \tau_1, t - \tau_2, t - \tau_3) \vdots \mathscr{E}_1 \mathscr{E}_2^* \mathscr{E}_3 \exp(i\omega_1\tau_1 - i\omega_2\tau_2 + i\omega_3\tau_3) \tag{8}$$

The natural description for frequency domain experiments is to express the electric field in terms of its Fourier transform $\mathbf{E}(\omega)$, yielding, for example, the third-order polarization

$$P_i^{(3)}(t) = \int_{-\infty}^{\infty} \int_{-\infty}^{\infty} \int_{-\infty}^{\infty} d\omega_1\, d\omega_2\, d\omega_3\, \chi_{ijkl}(\omega_1, \omega_2, \omega_3)$$

$$\times\, E_j(\omega_1)E_k(\omega_2)E_l(\omega_3)\exp[-i(\omega_1 + \omega_2 + \omega_3)t] \tag{9}$$

The nth-order susceptibilities $\chi^{(n)}$ are the Fourier transforms of the molecular response functions:

$$\chi^{(n)}(\omega_1 \cdots \omega_n) = \int_{-\infty}^{\infty} \cdots \int_{-\infty}^{\infty} dt_1 \cdots dt_n R^{(n)}(t_1 \cdots t_n) \exp\left(-i \sum_{j=1}^{n} \omega_j t_j\right) \quad (10)$$

In practice, the applied laser field consists of one or several pulses, each characterized by a pulse envelope in space and time, a mean frequency, and a propagation direction:

$$\mathbf{E}(\mathbf{r}, t) = \sum_j \tfrac{1}{2}\mathscr{E}(j; \mathbf{r}, t) \exp\{i(\omega_j t - \mathbf{k}_j \mathbf{r})\} + \text{c.c.} \quad (11)$$

The envelopes $\mathscr{E}(j; \mathbf{r}, t)$ often vary slowly in time on the scale of an optical period and the relevant molecular dynamics. Then the field is quasi-monochromatic,

$$\mathbf{E}(\mathbf{r}, \omega) = \sum_j \tfrac{1}{2}\{\mathscr{E}(j)\, \delta(\omega + \omega_j)e^{-i\mathbf{k}_j\mathbf{r}} + \mathscr{E}^*(j)\, \delta(\omega - \omega_j)e^{i\mathbf{k}_j\mathbf{r}}\} \quad (12)$$

and the convolution in Eq. (9), for example, reduces to sums of terms each corresponding to the interaction of three Fourier components of the field, such as

$$P_i^{(3)}(\mathbf{r}, t) = \chi_{ijkl}^{(3)}(\omega_1, -\omega_2, \omega_3)\mathscr{E}_j(1; \mathbf{r}, t)\mathscr{E}_k^*(2; \mathbf{r}, t)\mathscr{E}_l(3; \mathbf{r}, t)$$

$$\times \exp[i(\omega_1 - \omega_2 + \omega_3)t - i(\mathbf{k}_1 - \mathbf{k}_2 + \mathbf{k}_3)\cdot\mathbf{r}] \quad (13)$$

This polarization again has the form of a plane wave, with the frequency and wave vector the sum of the frequencies and wave vectors of the interacting field components. The sign of each contribution in these sums is positive for each $\mathscr{E}(j)$ and negative for each $\mathscr{E}^*(j)$. Thus knowledge of the product $\mathscr{E}(1)\mathscr{E}^*(2)\mathscr{E}(3)$ completely determines the whole expression of Eq. (13).

C. Nonlinear Dissipative Processes

The energy exchanged between the light beam and the molecular ensemble, per unit time and volume, is given by

$$\frac{dW}{dt} = \langle \mathbf{E} \cdot \dot{\mathbf{P}} \rangle \quad (14)$$

where the brackets indicate a time average over several cycles of the electric field. For monochromatic waves with amplitudes $\mathscr{E}$ and $\mathscr{P}$ and the same frequency ω, this average is

$$\frac{dW}{dt} = \tfrac{1}{2}\omega\text{Im}(\mathscr{E} \cdot \mathscr{P}) \quad (15)$$

As an example we consider two-photon absorption. The relevant nonlinear polarization at frequency ω is

$$\tfrac{1}{2}\mathscr{P}\exp(i\omega t) = \tfrac{1}{8}\chi^{(3)}(-\omega,\omega,\omega)\,\vdots\,\mathscr{E}\mathscr{E}\mathscr{E}^*\exp(i\omega t) \tag{16}$$

The energy absorbed through two-photon processes is therefore

$$\frac{dW}{dt} = \frac{8\pi^2\omega}{n^2c^2}I^2\mathrm{Im}(\chi^{(3)}) \tag{17}$$

where we have used $I = \mathscr{E}\mathscr{E}^*nc/8\pi$ for the intensity of the laser beam. In a rate equation description, two-photon absorption is often described through a cross section δ in units of cm^4 sec as

$$dn_p/dt = \delta NF^2 \tag{18}$$

where dn_p/dt is the number of photons absorbed per unit time, N the density of absorbing molecules, and $F = I/\hbar\omega$ the photon flux (McClain and Harris, 1978). Since $dW = dn_p\hbar\omega$, the relation between the cross section δ and the susceptibility $\chi^{(3)}$ is (Bechtel and Smith, 1976; Burris $et\ al.$, 1983):

$$\delta = \frac{8\pi^2\hbar\omega^2}{n^2c^2N}\mathrm{Im}(\chi^{(3)}) \tag{19}$$

Equation (19) can be used to measure two-photon cross sections through coherence experiments (Lotem and Lynch, 1976; Lynch and Lotem, 1977; Hochstrasser $et\ al.$, 1980). A similar relation is valid for Raman cross sections and can be deduced along the same lines (Bloembergen, 1965; Shen, 1974):

$$\frac{d^2\sigma}{d\omega_S d\Omega} = \frac{\hbar\omega_L\omega_S^3}{\pi c^4}\mathrm{Im}[\chi^{(3)}(\omega_L,\omega_S,-\omega_L)] \tag{20}$$

This kind of relation is quite general: all cross sections for absorption and stimulated emission can be related to the imaginary part of a susceptibility. Even the spontaneous emission can be obtained from this classical description when the black-body radiation field is taken as the "stimulating" field.

D. Optics of Parametric Processes

The temporal and spatial evolution of the generated light wave is described by the inhomogeneous Maxwell equations with the nonlinear polarization as source term:

$$\mathbf{V}\times\mathbf{V}\times\mathbf{E} + \frac{\varepsilon(\omega)}{c^2}\left(\frac{\partial^2}{\partial t^2}\right)\mathbf{E} = -\frac{4\pi}{c^2}\left(\frac{\partial^2}{\partial t^2}\right)\mathbf{P}^{NL} \tag{21}$$

$$\mathbf{V}\cdot[\varepsilon(\omega)\mathbf{E} + 4\pi\mathbf{P}^{NL}] = 0$$

We treat the complex dielectric tensor $\varepsilon(\omega) = 1 + 4\pi\chi^{(1)}(\omega)$ as a scalar, which is appropriate for isotropic media or beams travelling perpendicular to a principal direction in birefringent crystals. To account for linear absorption, we take the wave vectors of all waves to be complex:

$$\mathbf{K}_j = \hat{\mathbf{e}}_j(k_j - ia_j) \tag{22}$$

In this expression, $\hat{\mathbf{e}}_j$ is a unit vector perpendicular to the travelling wavefront, k_j is the usual wave vector in the medium with length $n\omega/c$, and a_j is half the linear absorption coefficient of the medium at frequency ω_j.

When the polarization wave is not attenuated through nonlinear depletion of the pump waves (i.e., for small conversion efficiencies), integration of Maxwell's equation with the slowly varying envelope method (Bloembergen, 1965; Butcher, 1965) yields for the amplitude of the generated wave

$$\mathcal{E}(L, t_{\mathrm{R}}) = \frac{2\pi\omega^2}{c^2 K_{\mathrm{S}}} \, \mathcal{P}(t_{\mathrm{R}}) \cdot \frac{\exp(i\Delta KL) - 1}{\Delta K} \tag{23}$$

In this expression L is the interaction length, $\Delta K = K_s - K_p$ is the phase mismatch, and K_s and K_p are the complex wave vectors of the signal and the polarization wave, respectively. The time argument t_{R} is a reduced time,

$$t_{\mathrm{R}} = t - \mathbf{k}_{\mathrm{S}} \cdot \mathbf{r}/\omega \tag{24}$$

which serves to effectively transform into a rest frame for the light pulses in time-domain processes. The instantaneous intensity of the generated light beam after an interaction length L is

$$I_{\mathrm{S}}(L, t_{\mathrm{R}}) = \frac{cn_{\mathrm{S}}}{8\pi} |\mathcal{E}(L, t_{\mathrm{R}}) \exp[i(\omega t - \mathbf{k}_{\mathrm{S}}\mathbf{r})]|^2$$

$$= \frac{n_{\mathrm{S}}\pi\omega^4}{2c^3(k_{\mathrm{S}}^2 + a_{\mathrm{S}}^2)} \cdot |\mathcal{P}^{\mathrm{NL}}(t_{\mathrm{R}})|^2 \cdot G(L) \tag{25}$$

Usually the signal is not resolved in time, but the integral is measured:

$$I_{\mathrm{S}}(L) = \frac{n_{\mathrm{S}}\pi\omega^4}{2c^3(k_{\mathrm{s}}^2 + a_{\mathrm{s}}^2)} G(L) \cdot \int_{-\infty}^{+\infty} dt_{\mathrm{R}} |\mathcal{P}^{\mathrm{NL}}(t_{\mathrm{R}})|^2 \tag{26}$$

The phase mismatch and absorption factor

$$G(L) = L^2 \frac{\sinh^2(\Delta aL/2) + \sin^2(\Delta kL/2)}{(\Delta aL/2)^2 + (\Delta kL/2)^2} e^{-(a_{\mathrm{s}} + a_{\mathrm{p}})L} \tag{27}$$

where $\Delta a = a_{\mathrm{s}} - a_{\mathrm{p}}$, is of particular interest in spectroscopic applications since it contains all information regarding the growth of the signal in propagating through the sample. It is necessary for estimating the optimum concentrations and interaction lengths. In the absence of absorption, $G(L)$ can increase as L^2. In the presence of absorption, an optimum exists for the interaction length, depending on a_s and a_p. Since a_s and a_p are both proportional to the concentration c of the molecular system, the signal can be optimized for a particular interaction length L through the choice of concentration. The interaction length itself is usually given through the confocal parameters of the beam-combining optics or the size of the sample cuvette. The signal depends on the concentration c through $a_s = \sigma_s c/2$ and $a_p = \sigma_p c/2$, where σ_s and σ_p are the molecular absorption cross sections for the signal and polarization waves. With phase matching, the signal is a function of the product Lc, which has the optimum value

$$(Lc)_{\text{opt}} = \frac{2\ln(\sigma_s/\sigma_p)}{\sigma_s - \sigma_p} \tag{28}$$

E. Molecular Theory of the Susceptibilities

To connect the observed macroscopic quantities to the molecular properties of interest, we write the macroscopic polarization as the ensemble average of the induced molecular dipoles,

$$\mathbf{P} = N\langle \boldsymbol{\mu} \rangle \tag{29}$$

where N is the number density of the molecules in the sample. The ensemble average is the trace of the product of the molecular dipole operator with the statistical (or density) operator ρ of the molecule:

$$\langle \boldsymbol{\mu} \rangle = \text{Tr}\{\rho \boldsymbol{\mu}\} \tag{30}$$

The density operator in turn obeys an equation of motion (Liouville equation) of the following form:

$$\dot{\rho} = \frac{i}{\hbar}[\rho, V] + \dot{\rho}^{\text{R}} \tag{31}$$

All operators are in the interaction picture, and the interaction operator $V(t)$ in the dipole approximation is

$$V(t) = -\boldsymbol{\mu}(t) \cdot \mathbf{E}(t) \tag{32}$$

where the matrix elements are $\mu_{ab}(t) = \mu_{ab}\exp(i\omega_{ab}t)$, $\omega_{ab} = (\varepsilon_a - \varepsilon_b)/\hbar$, and ε_i are the level energies of the molecule. The last term $\dot{\rho}^{\text{R}}$ in Eq. (31) is a phenomenological damping term accounting for the relaxation of the molecular excitations. Its particular form depends on the model for the partitioning of the molecular system into relevant states and bath states. In the

Markov approximation limit the decay is assumed to be exponential, and the matrix elements of $\dot{\rho}^{R}$ are

$$\dot{\rho}^{R}_{\mu\mu} = -\Gamma_{\mu\mu}\rho_{\mu\mu} + \sum_{\nu \neq \mu} \gamma_{\nu\mu}\rho_{\nu\nu} \tag{33}$$

$$\dot{\rho}^{R}_{\mu\nu} = -\Gamma_{\mu\nu}\rho_{\mu\nu} \tag{34}$$

Equation (33) describes the decay of the populations (diagonal elements of ρ) with rate constants $\Gamma_{\mu\mu}$ (inverse lifetimes T_1), as well as the feeding through the decay of other levels ν to the level μ with rate constants $\gamma_{\nu\mu}$. Equation (34) accounts for the decay of the off-diagonal elements (coherence loss) with the phase-relaxation rate $\Gamma_{\mu\nu} = 1/T_2(\mu\nu)$. This phase-relaxation rate is the mean of the population decay rates of the two levels plus a pure dephasing rate $\Gamma'_{\mu\nu}$:

$$\Gamma_{\mu\nu} = \tfrac{1}{2}(\Gamma_{\mu\mu} + \Gamma_{\nu\nu}) + \Gamma'_{\mu\nu} \tag{35}$$

The system and the bath are often chosen such that the latter vanishes under collision-free conditions, and in solids for low temperatures ($T \to 0$).

The system of coupled linear differential equations [Eq. (31)] can be decoupled in two steps. First the statistical operator is expanded in a Taylor series in powers of the interaction (i.e., electric field strength) by analogy to the expansion of the polarization:

$$\rho = \rho^{(0)} + \rho^{(1)} + \rho^{(2)} + \rho^{(3)} + \cdots \tag{36}$$

The second step is the neglect of the feeding constants $\gamma_{\nu\mu}$. This can be justified *a posteriori* when the perturbation theory expression for a particular matrix element of interest involves no excited-state populations, which is true in many important examples. The decoupled equations can then be integrated to yield (in the following we set $\hbar = 1$)

$$\rho^{(n+1)}_{\mu\nu} = ie^{-\Gamma_{\mu\nu}t} \int_{-\infty}^{t} dt' \, e^{\Gamma_{\mu\nu}t'} [\rho^{(n)}(t'), V(t')]_{\mu\nu} \tag{37}$$

The density matrix can now be iteratively calculated to any desired order. When the perturbation is a monochromatic wave in each step, the density matrix in each order will decompose into a sum of Fourier components at all combination frequencies of the ingoing fields. After taking the trace with the molecular dipole operator, the susceptibilities are recognized as the various Fourier coefficients:

$$\mathrm{Tr}\{\rho^{(n)}\boldsymbol{\mu}\} = \chi^{(n)} \cdot \mathbf{E}^{n} \tag{38}$$

There are, however, many time orderings and different pathways that contribute to the same Fourier component. For example, a Fourier component at frequency $\omega_4 = \omega_1 - \omega_2 + \omega_3$ is generated in third order by perturbations with the field components $\mathscr{E}_1$, $\mathscr{E}_2^*$, and $\mathscr{E}_3$ considered in $V(t)$, for which six different time orderings are possible. Furthermore, the interaction can be through the part ρV or the part $-V\rho$ of the commutator $[\rho, V]$ in Eq. (37). This is sometimes called evolution of the bra part or the ket part of the density operator. The interference between all these contributions often plays an important role in steady-state experiments, and diagrammatic methods have been developed to find all important terms (Bordé, 1976; Bordé and Bordé, 1978; Yee *et al.*, 1977; Yee and Gustafson, 1978; Druet *et al.*, 1978; Oudar and Shen, 1980). A method that is especially useful when molecular resonances are considered is shown in Fig. 1 (Dick and Hochstrasser, 1983a; Bozio *et al.*, 1983).

The figure shows the evolution of a particular diagram from left to right. The broken and full vertical arrows represent evolution of the ket and bra parts, respectively, in time ordering from left to right. The numbers at the bottom of the diagram below each arrow indicate the frequency of the field

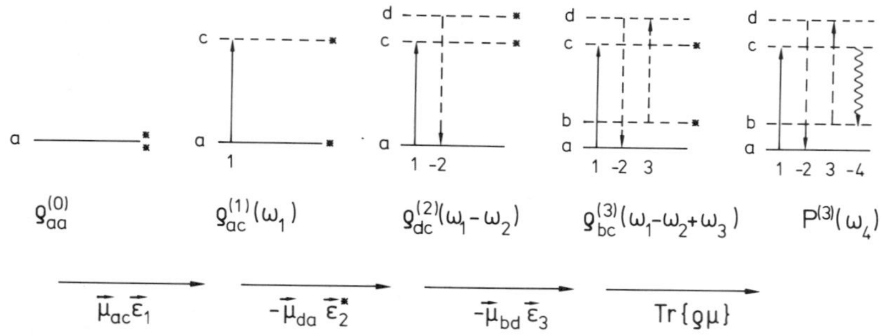

$$P^{(3)}(\omega_4)= \varrho_{aa}^0 \cdot \frac{\bar{\mu}_{ac}\bar{\varepsilon}_1}{(\omega_{ac}+\omega_1-i\Gamma_{ac})} \cdot \frac{-\bar{\mu}_{da}\bar{\varepsilon}_2^*}{(\omega_{dc}+\omega_1-\omega_2-i\Gamma_{dc})} \cdot \frac{-\bar{\mu}_{bd}\bar{\varepsilon}_3}{(\omega_{bc}+\omega_1-\omega_2+\omega_3-i\Gamma_{bc})} \cdot \bar{\mu}_{cb}$$

$$\underbrace{\qquad\qquad\qquad\qquad\qquad\qquad\qquad\qquad\qquad\qquad\qquad\qquad}$$

$$\chi_{ijkl}^{(3)}(\omega_3,-\omega_2,\omega_1)\,\varepsilon_j(1)\varepsilon_k^*(2)\varepsilon_l(3)$$

Fig. 1. Diagrammatic method for the calculation of nonlinear susceptibilities. As an example, a contribution to $\chi^{(3)}(\omega_3, -\omega_2, \omega_1)$ through a CSRS effect is shown. The development of the diagram through the orders of perturbation theory proceeds from left to right, the rightmost diagram being the final result. Each iteration adds a full or broken arrow to the diagram, representing evolution of the bra or the ket side of the density operator. The head of the latest full arrow and the tail of the latest broken arrow determine the indices of the current matrix element of the density operator. These are given in the second row in the figure, and the corresponding levels are marked by asterisks in the diagram. The third row gives the perturbation considered in each step coupling the connected levels. A minus sign indicates evolution of the ket (corresponding to a broken arrow in the diagram), and asterisk denotes the negative frequency component of the field (corresponding to a downward arrow). The formula represented by the diagram is given, where the five factors from left to right correspond to the five steps in building the diagram.

component. The arrow points upward when the positive frequency (i.e., photon creation) is involved, and downward if it is the negative frequency (i.e., the complex conjugate wave). The first arrow of each type starts from the level labeled "a," which represents the initial state. The full arrows point in the sense of the transition, whereas the broken arrows go in the opposite sense.

The corresponding mathematical expression starts with the population of the initial state, here $\rho_{aa}^{(0)}$. Each step in the perturbation expansion contributes a factor whose numerator is the coupling matrix element, and whose denominator is

$$\omega_{\mu\nu} + \sum_j \omega_j - i\Gamma_{\mu\nu} \tag{39}$$

Here $(\mu\nu)$ is the index pair of the density matrix element being calculated (marked with the asterisk), and the sum is over all frequencies involved in the process up to this point. The last dipole factor comes from the trace operation and corresponds to the curly arrow that closes the loop and represents the generated wave.

The same diagrammatic technique can be applied to obtain the response functions, by using delta-function pulses, as in Eq. (7), in the iterative calculation of the density matrix [Eq. (37)]. After taking the trace $\mathrm{Tr}(\rho\boldsymbol{\mu})$, the molecular expression for the polarization is of the form of Eq. (8), and the response function is easily identified. By analogy with the susceptibilities, a recipe can be given to obtain the response function directly from the corresponding diagram. The first term is again the population of the initial state, $\rho_{aa}^{(0)}$. Each iteration yields a factor

$$-i\mu_{xy}\exp[+(i\omega_{\mu'\nu'} + \Gamma_{\mu'\nu'})t_n - (i\omega_{\mu\nu} + \Gamma_{\mu\nu})t_n] \tag{40}$$

Here (xy) is the index pair of the transition dipole matrix element involved in the interaction with the field ε_n at time τ_n and $t_n = t - \tau_n$. The index pair $(\mu\nu)$ is the one of the required density matrix element, whereas the index pair $(\mu'\nu')$ refers to the density matrix element of the previous iteration step. The last factor is again the transition dipole element from the trace operation. For the diagram developed in Fig. 1, the response function is

$$R^{(3)}(t_1, t_2, t_3) = -i\rho_{aa}^{(0)}\mu_{ac}\mu_{da}\mu_{bd}\mu_{cb}$$

$$\times \exp\{-[(i\omega_{ac} + \Gamma_{ac})(t_1 - t_2)$$

$$+ (i\omega_{dc} + \Gamma_{dc})(t_2 - t_3) + i\omega_{bc} + \Gamma_{bc})t_3]\} \tag{41}$$

To obtain the correct susceptibility from the response function of Eq. (41), the Fourier transform must be performed with $\exp[i(\omega_1 t_1 - \omega_2 t_2 + \omega_3 t_3)]$.

The horizontal broken lines in Fig. 1 really refer to the complete set of eigenstates of the molecule, or virtual states having the photon energy, and the correct expression for the contribution to the susceptibility or the response function must be summed over all indices. It is clear that many different time orderings and combinations of bra and ket evolutions will contribute to the same Fourier component of ρ, but only a few diagrams have important contributions when resonances are considered.

As an example we discuss sum frequency mixing with two ingoing frequencies. The relevant Fourier component of ρ is $\rho^{(2)}(\omega_1 + \omega_2)$. Two time orderings are possible, namely $\mathscr{E}_1\mathscr{E}_2$ and $\mathscr{E}_2\mathscr{E}_1$. In each step the interaction can be on the bra or the ket side, making a total of eight diagrams. These are given in Fig. 2.

For a diagram to give any contribution at all, the starting level, marked by a dot in Fig. 2, must be populated. When this is the ground state of the molecule, all states that can act as intermediate states (broken lines) lie at higher energies. Diagrams involving arrows connecting the ground state with an even lower-lying intermediate state will always be far off-resonant, and can be neglected in the presence of other resonant diagrams. This neglect is mathematically equivalent to making the rotating wave approximation (RWA). It follows that only diagrams 1 and 5 in Fig. 2 need to be considered for a molecule in its

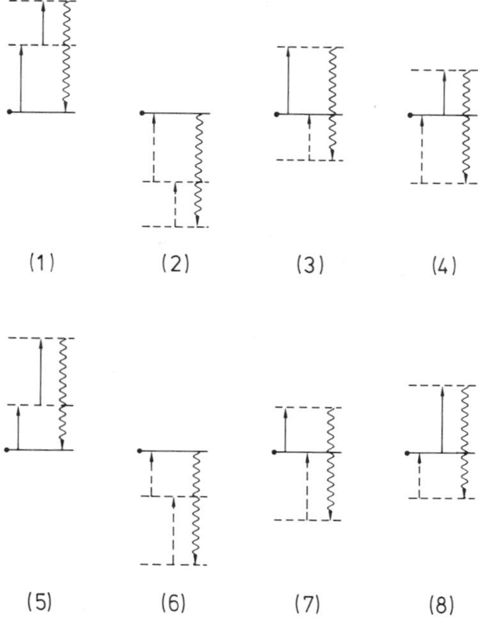

Fig. 2. The eight time-ordered diagrams for sum-frequency generation in second order, representing $P(\omega_1 + \omega_2)$. The initial state is marked by the dot.

ground state. Of course, diagrams like 3 and 4 can become important for processes starting from an excited state.

Extension of the method to third-order processes is straightforward, although the number of possible diagrams increases dramatically. For a Fourier component resulting from the mixing of three different frequencies, six time orderings and 48 diagrams arise. The Fourier component $2\omega_1 - \omega_2$ familiar from CARS (coherent anti-Stokes Raman scattering) still contains 24 diagrams, but 16 of them involve intermediate states below the starting level. The remaining eight are shown in Fig. 3. The introduction of resonance conditions will assign these to different physical processes. With $2\omega_1$ chosen in resonance with a transition $|a\rangle \to |d\rangle$, only diagrams 1 and 2 will contribute. Although at first glance diagrams 7 and 8 might seem important too, the

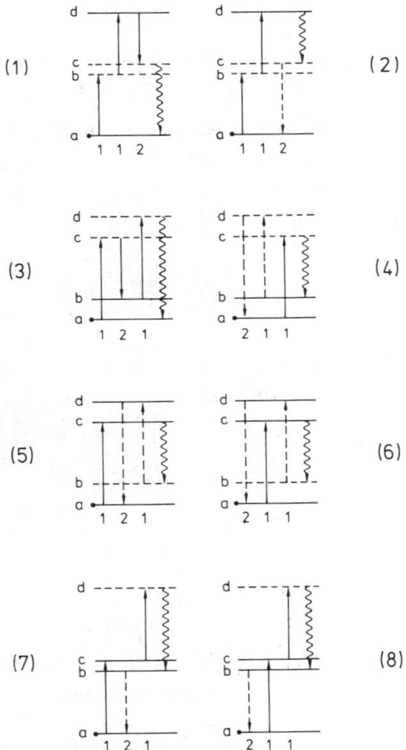

Fig. 3. Eight out of the possible 24 diagrams for $P(2\omega_1 - \omega_2)$. The other 16 diagrams can be neglected due to the rotating-wave approximation for molecules in their ground state (marked by the dot). Molecular resonances, indicated by the full horizontal lines at the energies of real states, allow these diagrams to be assigned to various spectroscopies: (1, 2) two-photon spectroscopy; (3) CARS of a ground-state Raman transition; (4) CSRS of a ground-state Raman transition; (5, 6) CSRS of an excited-state Raman transition; (7, 8) CARS of an excited-state Raman transition.

corresponding $\chi^{(3)}$ shows no resonance denominator related to the two-photon resonance. This is because the path through the perturbation theory never leads to a coherence ρ_{ad} for these diagrams. [The paths are $\rho_{ac}^{(1)}$ to $\rho_{bc}^{(2)}$ to $\rho_{bd}^{(3)}$ and $\rho_{ba}^{(1)}$ to $\rho_{bc}^{(2)}$ to $\rho_{bd}^{(3)}$]. Of particular interest besides two-photon resonances are Raman resonances in ground and excited states. Ground-state Raman resonances are contained in diagram 3 for $\omega_1 > \omega_2$ and diagram 4 for $\omega_1 < \omega_2$. These are the classical CARS and CSRS (coherent Stokes Raman scattering). The excited-state Raman resonances require that the ingoing frequencies are also each resonant with a one-photon transition. For the CSRS configuration (diagrams 5 and 6), a molecular vibrational level will always lie close to the position for the intermediate level $|b\rangle$, making these diagrams in effect fully resonant. Interference between these two diagrams results in the extra resonances.

F. Discussion of Examples

The susceptibilities can be classified according to the number of resonances. This can lead to considerable simplification of the mathematical expressions. As an example we take the CARS susceptibility for a molecule in its ground state in the rotating-wave approximation (diagram 3 of Fig. 3):

$$\chi_{CARS}^{(3)} = \sum_{b,c,d} \frac{\mu_{ac}\mu_{cb}\mu_{bd}\mu_{da}}{(\omega_{ac} + \omega_1 - i\Gamma_{ac})(\omega_{ab} + \omega_1 - \omega_2 - i\Gamma_{ab})(\omega_{ad} + 2\omega_1 - \omega_2 - i\Gamma_{ad})} \tag{42}$$

When ω_1 and ω_2 are both far from any electronic resonance of the molecule, only the second denominator will lead to resonances for those states $|b\rangle$ for which $\omega_{ab} + \omega_1 - \omega_2 = 0$ in the range over which ω_1 and ω_2 are scanned in the experiment. In this case we can write

$$\chi_{CARS}^{(3)} = \sum_{b}' \frac{R_{ab}}{\omega_{ab} + \omega_1 - \omega_2 - i\Gamma_{ab}} + \chi_{NR}^{(3)} \tag{43}$$

The prime at the summation symbol indicates that the sum is only over resonant states $|b\rangle$. All other terms are incorporated in the nonresonant susceptibility $\chi_{NR}^{(3)}$, which is often constant in the frequency range of interest. The amplitudes R_{ab} are related to the Raman transition polarizabilities $\alpha(\omega)$:

$$R_{ab} = \alpha_{ab}(\omega_1)\alpha_{ba}(2\omega_1 - \omega_2) \tag{44}$$

$$\alpha_{ab}(\omega) = \sum_{c} \frac{\mu_{ac}\mu_{cb}}{\omega_{ac} + \omega - i\Gamma_{ac}}$$

The damping parameters Γ can be neglected in all nonresonant denominators, making χ_{NR} and the Raman amplitudes real. Other single resonant susceptibilities can be contracted in the same way. For example, the two-photon resonant susceptibility resulting from the diagrams 1 and 2 of Fig. 3 can be written as

$$\chi^{(3)}_{\text{TPA}} = {\sum_d}' \frac{T_{ad}}{\omega_{ad} + 2\omega_1 - i\Gamma_{ad}} + \chi^{(3)}_{\text{NR}}$$

$$T_{ad} = \alpha_{ad}(\omega_1)[\alpha_{da}(\omega_2) + \alpha_{da}(2\omega_1 - \omega_2)]$$

(45)

Difference-frequency generation is an example of a doubly resonant process. The two relevant diagrams are shown in Fig. 4. With the method outlined above, we obtain for the resonant part of the response functions

$$R_1(t_2, t_1) = -\mu_{ac}\mu_{ba}\mu_{cb} \exp[-(i\omega_{ac} + \Gamma_{ac})(t_2 - t_1) - (i\omega_{bc} + \Gamma_{bc})t_1] \quad (46)$$

$$R_2(t_1, t_2) = -\mu_{ba}\mu_{ac}\mu_{cb} \exp[-(i\omega_{ba} + \Gamma_{ba})(t_1 - t_2) - (i\omega_{bc} + \Gamma_{bc})t_2] \quad (47)$$

Excitation by very short light pulses yields a nonlinear polarization given by these response functions, and the integrated signal at the detector is

$$I_1 = \int_{\tau_1}^{\infty} |R_1(t - \tau_2, t - \tau_1)|^2 \, dt = \frac{\mu_{ba}^2 \mu_{ac}^2 \mu_{cb}^2}{2\Gamma_{bc}} \exp[-2\Gamma_{ac}(\tau_1 - \tau_2)] \quad (48)$$

$$I_2 = \int_{\tau_2}^{\infty} |R_2(t - \tau_1, t - \tau_2)|^2 \, dt = \frac{\mu_{ba}^2 \mu_{ac}^2 \mu_{cb}^2}{2\Gamma_{bc}} \exp[-2\Gamma_{ab}(\tau_2 - \tau_1)] \quad (49)$$

Different time ordering of the two light pulses will thus measure different relaxation times of the molecule. The susceptibilities corresponding to the two diagrams are

$$\chi_1 = \frac{-\mu_{ac}\mu_{ba}\mu_{cb}}{(\omega_{ac} + \omega_2 - i\Gamma_{ac})(\omega_{bc} + \omega_2 - \omega_1 - i\Gamma_{bc})} \tag{50}$$

$$\chi_2 = \frac{-\mu_{ac}\mu_{ba}\mu_{cb}}{(\omega_{ba} - \omega_1 - i\Gamma_{ab})(\omega_{bc} + \omega_2 - \omega_1 - i\Gamma_{bc})} \tag{51}$$

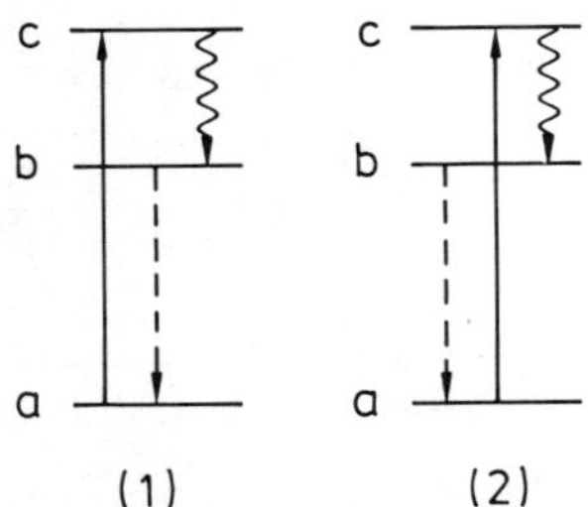

Fig. 4.   The two time-ordered diagrams describing difference-frequency generation.

They can be obtained by Fourier-transforming Eqs. (46) and (47), or directly from the diagrams. In a frequency-domain experiment, the time ordering of the interactions cannot be distinguished, and the effective susceptibility is the sum of χ_1 and χ_2:

$$\chi_{\text{eff}} = \chi_1 + \chi_2 = \frac{\mu_{ac}\mu_{ba}\mu_{cb}}{(\omega_{ac} + \omega_2 - i\Gamma_{ac})(\omega_{ab} + \omega_1 + i\Gamma_{ab})} \tag{52}$$

$$\times \left(1 - i\frac{\Gamma'_{ab} + \Gamma'_{ac} - \Gamma'_{bc}}{\omega_{bc} + \omega_2 - \omega_1 - i\Gamma_{bc}}\right)$$

The result of the interference of both diagrams is the so-called DICE resonance (Andrews and Hochstrasser, 1981a,b; Andrews *et al.*, 1981) appearing in the large parentheses. The numerator of this resonance contains only pure dephasing rates, and thus the amplitude of this resonance is a measure of pure dephasing events.

G. Inhomogeneous Broadening and Line Narrowing

The susceptibilities or response functions calculated up to this point refer to a single molecule or a homogeneous ensemble of noninteracting molecules. In inhomogeneous systems, different sets of molecules have different transition energies, and the total susceptibility is the ensemble average. For a single resonance, the average is

$$\langle\chi\rangle = \int_{-\infty}^{+\infty} dx\, g(x)\frac{A}{\omega_A + x + \omega - i\Gamma} \tag{53}$$

where ω_A is the mean transition frequency of the ensemble, x the frequency shift for a particular molecule, and $g(x)$ the probability distribution function of the inhomogeneous distribution. When the inhomogeneous distribution is the result of a large number of independent perturbations (e.g., due to different environments in a solid), the distribution probability $g(x)$ is Gaussian. This is also the case for the Doppler distribution in gas phase. In this case,

$$g(x) = (2\pi\sigma^2)^{-1/2} \exp(-x^2/2\sigma^2) \tag{54}$$

$$\langle\chi\rangle_\sigma = i(\pi/2)^{1/2}\sigma^{-1}AW\left[\frac{i\Gamma - \omega_A - \omega}{(2\sigma)^{1/2}}\right] \tag{55}$$

where $W(z)$ is the complex error function. Its width is mainly determined by the larger of the two widths Γ and σ, the homogeneous and inhomogeneous widths. The main effect is immediately visible when $g(x)$ is approximated by a

Lorentzian:

$$g(x) = (\sigma/\pi)(\sigma^2 + x^2)^{-1} \tag{56}$$

$$\langle \chi \rangle_{\mathrm{L}} = \frac{A}{\omega_A + \omega - i(\Gamma + \sigma)} \tag{57}$$

The linewidth of the averaged susceptibility is the sum of the homogeneous and the inhomogeneous width. When the inhomogeneous width is dominant, the singly resonant susceptibility cannot be used to obtain homogeneous parameters of the system. In multiply resonant systems, however, the average over broad inhomogeneous distributions can lead to homogeneous resonance lines. To understand these so-called line-narrowing effects, let us consider a doubly resonant susceptibility,

$$\chi(x, y) = \frac{A}{(\omega_a + \omega_1 + x - i\Gamma_1)(\omega_b + \omega_2 + y - i\Gamma_2)} = \frac{A}{(\Omega_a + x)(\Omega_b + y)} \tag{58}$$

where ω_a and ω_b are the center frequencies of the transitions, and x and y the frequency shifts for a particular molecule. The distribution of the frequency shifts for the ensemble of molecules is described by a two-dimensional distribution function $g(x, y)$. When the distributions of x and y are uncorrelated, the distribution function factors into $g(x, y) = g_1(x)g_2(y)$, and the average

$$\langle \chi \rangle = \int\!\!\int dx\, dy\, \chi(x, y)g(x, y) \tag{59}$$

breaks up into the product of two terms like Eq. (54), leading to broad inhomogeneous resonances. In the other extreme of complete correlation between the distributions,

$$g(x, y) = \tilde{g}(x)\,\delta(y - \alpha x) \tag{60}$$

the integral reduces to

$$\langle \chi \rangle = \int dx\, \tilde{g}(x)\frac{A}{(\Omega_a + x)(\Omega_b + \alpha x)} \tag{61}$$

When $\tilde{g}(x)$ is approximated by a Lorentzian, Eq. (62) can be solved by contour integration. When α is positive there is only one pole, at $-i\sigma$, lying below the real axis, so that the resulting $\langle \chi \rangle$ is

$$\langle \chi \rangle = \frac{A}{(\Omega_a - i\sigma)(\Omega_b - i\alpha\sigma)} \tag{62}$$

This contains two broad resonances. However, when α is negative, a second term arises and the susceptibility becomes

$$\langle \chi \rangle = \frac{A}{\Omega_b - \alpha\Omega_a} \cdot \left[\frac{1}{\Omega_a - i\sigma} - \frac{\alpha}{\Omega_b + i\alpha\sigma} \right] \quad ; \alpha < 0 \qquad (63)$$

The susceptibility now has a resonance denominator with only homogeneous linewidth parameters. The average can also be performed with a gaussian distribution function, and the numerical evaluation of the complex error function shows the same effect (Dick and Hochstrasser, 1983a). Thus, the line-narrowing properties are intimately connected with the correlations within the inhomogeneous lines for each of the transitions involved in the overall process (i.e., each of the resonances).

H. Strong Light Fields

The perturbative description of the interaction between light fields and molecular systems breaks down when very strong light fields are resonant with molecular transitions. Since our interest is in these resonant situations, a careful analysis of the applicability of perturbation theory is in order. For the two-level system, the exact solution of the Liouville equation [Eq. (31)] within the rotating-wave approximation is well known and has been discussed many times. [For a review, see Allen and Eberly (1975) or Schmalz and Flygare (1978).] The exact result is well represented by the perturbation theory expression as long as

$$\Omega = \mu_{12}E/2\hbar \ll (\Gamma_{12}^2 + \Delta^2)^{1/2} \qquad (64)$$

where μ_{12} is the transition dipole, E the field amplitude, Γ_{12} the transverse relaxation rate, Δ the detuning from resonance, and Ω the Rabi frequency (Rabi, 1937). For large field strengths that violate Eq. (64), the perturbation theory is no longer applicable. The exact solution describes power broadening, saturation, ac Stark effects, and optical nutation.

Strong field effects are also expected in nonlinear spectroscopic experiments in multilevel systems if one or more of the ingoing laser fields violates Eq. (64) for a particular transition. The simplest case is an N-level system in full resonance with $N - 1$ laser fields. In this case, a unitary transformation will remove all rapidly time-dependent terms (Dick and Hochstrasser, 1983b). In the steady state, a simple linear equation system yields the density operator of the molecular system. The latter can be used to calculate parametric processes (Dick and Hochstrasser, 1983b), as well as stimulated or spontaneous dissipative processes (Dick and Hochstrasser, 1983c, 1984b). Two interesting new effects are predicted:

1.   In parametric processes, a splitting of resonances without power-broadening is possible. For example, in a CARS experiment where the interaction of the pump beam with an electronic transition is strong, the Rabi frequency of this interaction will show up as a splitting of the CARS resonance, whereas the width of the CARS resonance will not be affected. This effect could be used to measure transition dipoles.

2.   The extra resonances induced through pure dephasing processes in the framework of perturbation theory will appear as power-induced extra resonances, even in the absence of pure dephasing.

IV. SELECTED EXAMPLES

A. Introduction

Since tunable dye lasers became available over 10 years ago, a great variety of nonlinear spectroscopic techniques has been developed and applied to molecular condensed matter. The first high-resolution two-photon spectra of molecular crystals were obtained by Hochstrasser *et al.* (1973a,b, 1974), who recorded the spectra of benzene, naphthalene, and biphenyl at liquid-helium temperatures by monitoring the ultraviolet (UV) fluorescence of these materials subsequent to two-photon absorption from a pulsed dye laser. The increase of spectroscopic information obtained in these experiments is analogous to the importance of Raman versus infrared for vibrational spectroscopy. In both cases different selection rules make new transitions observable, and six tensor versus three vector components probe in more detail the anistropy of a molecule. A technical advantage is that bulk crystals can be used because the materials are transparent at the incident laser frequency, yet only a small volume is probed, and these experiments are therefore less sensitive to strain, which is produced more readily in the very thin samples necessary for linear absorption experiments.

As lasers developed, more demanding experiments could be performed: coherent transients such as photon echos, optical nutation, and optical free induction decay were observed and used to probe the dynamics of molecular systems at low temperatures (for reviews, see Hesselink and Wiersma, 1983; Burns *et al.*, 1983). In the following we shall discuss experiments in which a new frequency component is produced in a parametric process: three- or four-wave mixing, such as sum- and difference-frequency generation via $\chi^{(2)}$, or CARS and CSRS via $\chi^{(3)}$. The spectroscopic information in these experiments is obtained by monitoring the intensity of the generated beam at the new frequency component as a function of the frequencies of the incident beams. In addition to the spectroscopic information, these experiments yield precise determinations of the nonresonant nonlinear susceptibilities and thus measure important material constants (nonlinear refractive indices). In molecular solids, four-wave mixing is also the most precise method to evaluate two-photon cross-sections of individual vibronic transitions. In these frequency-domain experiments, information about the dynamics is obtained from an analysis of the lineshape and width. This dynamical information can equally well be obtained in time-domain experiments, where the coherence decay is measured directly by monitoring the intensity of the generated signal as a function of the delay between the laser pulses. These measurements are superior for long coherence decay times or when a narrow level structure can be resolved from the resulting beat pattern of the coherence decay.

The choice of examples given below is necessarily arbitrary and is not intended as an exhaustive review. The examples are arranged according to the number of molecular resonance conditions that are simultaneously fullfilled: the case of no resonances is the focus of most of the other chapters in this book and will not be discussed here.

B. Single Resonances

Singly resonant contributions to the susceptibility are those for which only one resonance condition is nearly fullfilled. These resonant contributions occur when the difference or the sum of the frequencies of two incident fields equals the energy difference of a pair of levels connected by a Raman or a two-photon absorption process, or when the frequencies of either the incident or of the generated fields resonate in a one-photon process. At least one of the two levels coupled by the light field must be populated in order to observe such resonances. A system may have any number of singly resonant contributions and the susceptibility has the general form

$$\chi = \langle \chi_{\mathrm{NR}} + \Sigma A_{ab}/(\omega_{ab} + \{\omega_l\} \pm i\Gamma_{ab}) \rangle \tag{65}$$

such as specified in Eqs. (43) and (45) for Raman or two photon resonances. The term $\{\omega_l\}$ stands for each of the combinations of frequencies satisfying $\omega_{ab} + \{\omega_l\} = 0$, such as $\{\omega_l\} = \omega_1 - \omega_2$ for Raman resonances and $\{\omega_l\} = 2\omega_1$ for two-photon resonances; the sign of $i\Gamma$, which is determined by the specific resonant process, can be obtained without calculation, by inspection of diagrams such as were explained in Section III,E. The angle brackets indicate that such singly resonant contributions are to be summed over all species present in a unit volume of the sample. These species may be different molecules or the same molecules in different environments or in different electronic, vibrational, or rotational (phonon) states. The term χ_{NR} is often a real quantity representing the contribution of all nonresonant transitions for which the frequency mismatch is much larger than the damping parameter.

In four-wave mixing experiments, the intensity of the coherently generated light is measured as a function of $\{\omega_l\}$. The light intensity is proportional to the square of the nonlinear source polarization (see Section III.D): thus it measures $|\chi^{(3)}|^2$, which is the sum of the squares of the real and imaginary parts of Eq. (65). For an isolated resonance, the signal will show a maximum and a minimum when

$$\omega_{ab} + \{\omega_l\} = \pm [(A_{ab}/2\chi_{NR})^2 + \Gamma_{ab}^2]^{1/2} - A_{ab}/2\chi_{NR} \tag{66}$$

The variation with frequency of a susceptibility of this type thus leads to the determination of resonance frequencies ω_{ab}, the corresponding damping parameter Γ_{ab}, and the amplitude of the resonance measured with respect to the nonresonant background A_{ab}/χ_{NR}.

C. Raman and Two-Photon Resonances in Four-Wave Mixing

Many experimental studies of singly resonant responses have involved Raman and two-photon resonances using $\chi^{(3)}$, the lowest-order susceptibility to which these resonances contribute. Four-wave mixing, arising from two different input frequencies ω_1 and ω_2 and monitored by the intensity of light generated by the induced polarization at $\omega_3 = 2\omega_1 - \omega_2$, has emerged as a versatile method of measuring singly resonant $\chi^{(3)}$ in gases, liquids, and solids. The frequencies of all light beams involved in this situation are similar, and phase matching is therefore easy to achieve in all media, yet the directional signal at $2\omega_1 - \omega_2$ is readily isolated. The generated beam can be chosen so that $\omega_3 > \omega_1 > \omega_2$ or that $\omega_3 < \omega_1 < \omega_2$. These Raman resonant contributions were given the acronyms CARS and CSRS, for coherent anti-Stokes (or Stokes) Raman scattering/spectroscopy. When χ_{NR} is real, both methods lead to the same spectral shapes for a given resonant level pair, but when two-photon and Raman resonances interfere, the CARS and CSRS spectra will differ because the signs of the damping parameters in Eq. (65) are different for CARS and CSRS.

In spontaneous Raman spectroscopy with conventional monochromators, the spectral resolution is about 0.2 cm^{-1}. Improvements can be accomplished by means of elaborate interferometric methods. In contrast, the spectral resolution in CARS is determined by the spectral bandwidth of the lasers. It was therefore natural to use CARS for high-resolution gas-phase Raman spectroscopy and to probe the chemical composition and the temperature distribution within flames (Regnier and Taran, 1973; Regnier *et al.*, 1974; Moya *et al.*, 1975). The generation of light at the anti-Stokes frequency is resonance-enhanced when $\omega_1 - \omega_2$ matches a vibrational frequency and also when there are two-photon resonances. It follows from Eq. (43) and Eq. (45) that the polarization in a medium displaying both Raman and two-photon effects takes the form

$$P^{(3)} = \langle \chi_{\text{NR}} + R_{ab}/(\omega_{ab} + \omega_1 - \omega_2 - i\Gamma_{ab}) + T_{ad}/(\omega_{ad} + 2\omega_1 - i\Gamma_{ad})\rangle \quad (67)$$

The amplitudes R_{ab} and T_{ad} of the Raman and two-photon resonances, respectively, are directly related to Raman scattering and two-photon absorption cross sections and are readily calibrated in these experiments. The comparatively low probability of two-photon absorption processes has resulted in the development of many indirect detection methods, but it is often not practical to use such techniques to determine absolute cross sections. Absolute Raman cross sections, on the other hand, have been measured for a number of molecules and, as Raman resonances are fairly narrow even at room temperatures, have been used in organic liquids and solids to determine $\chi^{(3)}$ (Levenson and Bloembergen, 1974a,b; Hochstrasser *et al.*, 1980). This calibration is in turn used to evaluate the two-photon tensor T_{ad} of individual vibronic lines. In condensed phases the vibrational structure of electronic transitions can only be resolved at low temperatures, and two-photon cross sections of individual vibronic bands can thus only be measured under these conditions. Measurements on single crystals of benzene, naphthalene, and biphenyl are discussed in detail by Hochstrasser *et al.* (1980). The spectra of neat benzene crystals at 1.6 K are shown in Fig. 5: the two-photon fluorescence excitation spectrum obtained by Hochstrasser *et al.* (1973a,b, 1974) is shown together with four wave-mixing spectra for the spectral region around the strongest two-photon transition corresponding to the excitation of a b_{2u} nuclear displacement on the $^1B_{2u}$ excited state surface, the overall two-photon process thus corresponding to a transition between two A_g states. The frequency difference $\omega_1 - \omega_2$ was chosen to be close to the frequency of skeletal modes (around 1600 cm^{-1}; upper trace) and of C—H stretch modes (around 3050 cm^{-1}; lower trace). The continuous line is a fitted spectrum calculated according to Eq. (67). In all three materials it was found that the magnitude of the two-photon tensor of the strongest transitions was similar to the corresponding values for the stronger Raman transitions.

More important than these calibrations of optical constants and transition strengths of single crystals at low temperatures is the fact that, because the coherent signals in these experiments can easily be measured over an extremely large dynamic range, line shapes can be accurately determined with narrow band lasers. Figure 6 shows the CARS spectrum of the pure naphthalene crystal (DeCola *et al.*, 1980b). In this case the stronger Raman

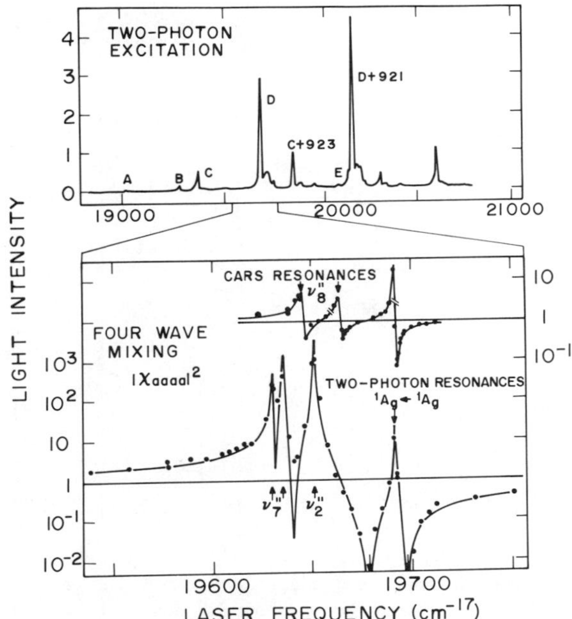

Fig. 5. Two-photon fluorescence (top) and four-wave mixing spectra (bottom) of benzene crystals at 1.6 K. The transition marked D in the top spectrum corresponds to the $14_0^1 (A_g)$ band of the $^1B_{2u} \leftarrow {}^1A_g$ electronic transition and is the strongest vibronic origin of the two photon spectrum. The same transition is seen in the bottom spectra, together with Raman resonances of ground-state vibrational modes; the frequency of the ω_1 laser is scanned while ω_2 is chosen such that $\omega_1 - \omega_2$ is in the region of C—C and C—H stretch modes, respectively, in the upper and lower traces.

resonance could be fitted to a single Lorentzian line with a Γ of 0.03 cm^{-1} corresponding to an exponential decay constant of 90 psec.

A fundamental issue in the analysis of line shapes in the condensed phase is the separation of inhomogeneous and homogeneous contributions. In the case of the pure crystal of naphthalene, the rapid delocalization of the 1385-cm^{-1} excitation averages out the inhomogeneous frequency distribution. This effect for excitons is analogous to the motional narrowing better known in magnetic resonance. The transition for a static inhomogeneous gaussian distribution to a motionally narrowed Lorentzian line is described in a model developed by Kubo (Kubo and Tomita, 1954; Kubo, 1969) in which the transition frequency is stochastically modulated to yield a response function given by

$$R(t) = \exp\{-\sigma^2\tau_c^2[\exp(-t/\tau_c) - 1 + t/\tau_c]\} \tag{68}$$

Here σ is the gaussian parameter of the frequency distribution and τ_c the correlation time of the stochastic jumps. For a crystal excitation in the

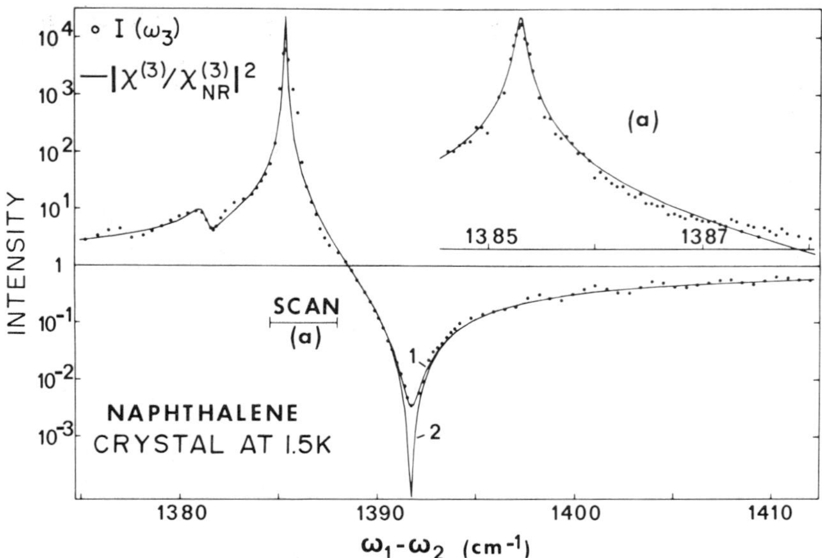

Fig. 6. CARS spectra of a naphthalene single crystal at 1.5 K. The directions of propagation of all beams are nearly perpendicular to the *ab* plane and all polarization are parallel to the *a* axis. The continuous trace was calculated according to Eq. (66) with $\chi_{NR} = |\chi_{NR}|(1 - i\varepsilon)$ where $\varepsilon = 0.05$ in curve 1 and $\varepsilon = 0$ in curve (2); $\omega_1 = 16{,}994$ cm^{-1}.

restricted Frenkel limit, τ_c corresponds to the jump time of a localized excitation between neighboring sites and σ is the gaussian width of the inhomogeneous distribution of site energies, as would be measured, for example, in a dilute mixed crystal. When the bandwidth β is small (i.e., τ_c very long), the coherence will decay on a time scale short compared with τ_c (provided that $\sigma\tau_c \gg 1$): in this limit, $R(t) \approx \exp(-\sigma^2 t^2/2)$, and in the frequency domain a gaussian line with the full inhomogeneous width σ is observed. When, on the other hand, $\sigma\tau_c \ll 1$, the decay becomes $R(t) \approx \exp(-\sigma^2 t/\beta)$, corresponding to a Lorentzian line of width σ^2/β: the exciton motion thus reduces the line width by a factor σ/β. A discussion of these points more specifically directed at exciton systems was given by Abram and Hochstrasser (1979). In low-temperature single crystals, the inhomogeneous contribution becomes negligible compared with the lifetime contribution to the line width as a result of this narrowing.

Dynamics of vibrational state relaxation in crystals can also be studied in time-domain experiments in which ω_1 and ω_2 first create the coherence between the $v = 0$ and $v = 1$ levels, and this coherence is probed at a later time by an ω_1 beam. A measurement of the decay of the light field generated at frequency $\omega_3 = 2\omega_1 - \omega_2$ yields the coherence decay parameter Γ for the two-level system. If in time-domain experiments the light intensity is measured, the observed decay constant is expected to be 2Γ. In the case of the naphthalene crystal 1385-cm^{-1} band, time-domain experiments yielded an exponential decay constant in excellent agreement with the CARS value (Hesp and Wiersma, 1980). Similar comparisons now exist for benzene modes (Ho *et al.*, 1981; Trout *et al.*, 1984). In fact, both time- and frequency-domain experiments measure the same response of the system. This is readily seen by inspection of the form of the CARS response function, R_{CARS}:

$$R_{\mathrm{CARS}} = \sum_{a,v,c,d} \rho_a \exp\{(i\omega_{ac} + \Gamma_{ac})(\tau_1 - \tau_2)$$
$$+ (i\omega_{av} + \Gamma_{av})(\tau_2 - \tau_3) + (i\omega_{ad} + \Gamma_{ad})(\tau_3 - t)\} \qquad (69)$$

In the conventional time resolved CARS experiment, $\tau_1 = \tau_2$ (the first two interactions come from pulses that are centered at the same instant). The vibrational levels of the ground state are labelled v. When the fields E_1 and E_2 are in the transparent regime and E_3 is delayed by τ, the CARS signal for nearly δ-function pulses has the form

$$I_{\mathrm{CARS}}(\tau) \sim \mathrm{const}\left|\sum_{a,v} \rho_a \exp(i\omega_{va} - \Gamma_{av})\tau\right|^2 \qquad (69b)$$

When the pulses have a finite spectral bandwidth, specific vibrational resonances can be excited, each having the asymptotic intensity decay function $\exp(-2\Gamma_{av}\tau)$. The effect of intermediate pulsewidths is obtained directly from Eqs. (6) and (69). The signal $I_{\mathrm{CARS}}(\tau)$ versus τ is a Fourier transform Raman spectrum. Note that the Fourier transform of Eq. (69), using Eq. (10), is given by Eq. (42). Only in the limit when the duration of the light pulses is much shorter than the coherence decay or when the frequency bandwidth of the laser is much smaller than the width of the resonance are the time and frequency CARS responses simply transforms of one another. In real experiments these conditions are frequently not met, and the spectral–temporal properties of the laser fields have to be convoluted with the system response via Eq. (6) or Eq. (9) in order to simulate properly the observed signals and to extract the physical meaningful parameters (Ho *et al.*, 1983).

Benzene is one of the best understood organic crystals. The 991-cm^{-1} ring-stretching vibration forms an exciton band in the crystal that can be excited in a Raman transition to its lowest-energy Davydov component by proper choice of the light polarization. Figure 7 shows measurements of the coherence decay by picosecond time-resolved CARS for this mode, as well as frequency-domain measurements in benzene crystals of natural isotopic composition: the decay times determined in both types of experiments agree with the error limits.

The study of vibrational linewidths in low-temperature molecular crystals by high-resolution CARS is only a few years old, but a clear picture of certain

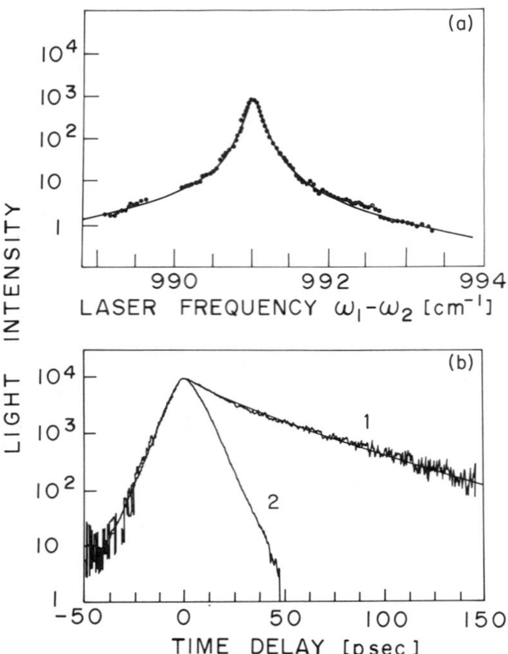

Fig. 7.   (a) Frequency and (b) time domain (curve 1) CARS spectra of the v_1 991-cm^{-1} mode of benzene single crystals of natural isotopic composition. Curve (2) is an instrument function obtained from liquid benzene. The linewidth parameter (0.067 cm^{-1}) obtained from the top spectrum agrees within experimental error with the decay time (40 psec) obtained from the bottom curve.

aspects of coherence decay in these systems begins to emerge. As the temperature of a crystal is lowered, the vibrational linewidth decreases until (around 10 K) it becomes close to a temperature-independent limit. The residual ($T = 0$) linewidth is due to dephasing or energy transfer. From a time-domain point of view, dephasing processes represent mechanisms by which initially well-defined phase relations among the excited oscillators in the wavepacket are randomized by strain fields, or by defect and impurity scattering. This can be pictured as the evolution of a relatively smooth Fourier-limited initial excitation wavepacket into a structure characterized by a correlation length inversely proportional to the width of the distribution of k states into which the wavepacket has evolved. In energy-transfer processes, vibrational amplitude diminishes evenly at all excited sites of the crystal, preserving all initially present phase relationships. This kind of decay is accompanied by the spontaneous emission of phonons and corresponds to a loss of vibrational population from the initially excited mode (T_1 process). In a pure crystal, from which all chemical and isotopic impurities have been removed, only strain-induced dephasing and lifetime broadening due to spontaneous emission of phonons contribute to the vibrational linewidth.

An important goal of these nonlinear studies of crystals is to learn about molecular relaxation processes. The population decay rates for benzene were found to increase with the total vibrational energy content for benzene as shown in Fig. 8. The benzene crystal is sufficiently harmonic that $\Delta v = \pm 1$ selection rules were expected to determine the relaxation pathways involving both the internal and the external, relative motion, degrees of freedom. An explanation of this trend was recently given in terms of the variations of the anharmonicities of the benzene molecular modes. The internal anharmonicities of the benzene modes are expected to increase with vibrational frequency. At low vibrational energy, say less than 1000 cm^{-1}, there is a harmonic region within which the existence of $\Delta v = \pm 1$ possibilities is not sufficient for a rapid relaxation. In the higher-energy, or "free-access," region, coupling may occur to many of the levels that are separated by less than one lattice-vibrational quantum, as a result of the increased molecular anharmonicities and Fermi resonances. At a certain state density still in the region of the fundamentals, the molecular motions might be sufficiently mixed to allow each mode to relax into many channels while still maintaining the $\Delta v = \pm 1$ selection rule for lattice modes. Internal mode mixing of this nature was also invoked to describe the relaxation of CH stretching modes in liquids (Fendt *et al.*, 1981) specifically in that case in terms of their coupling to CH bending motions.

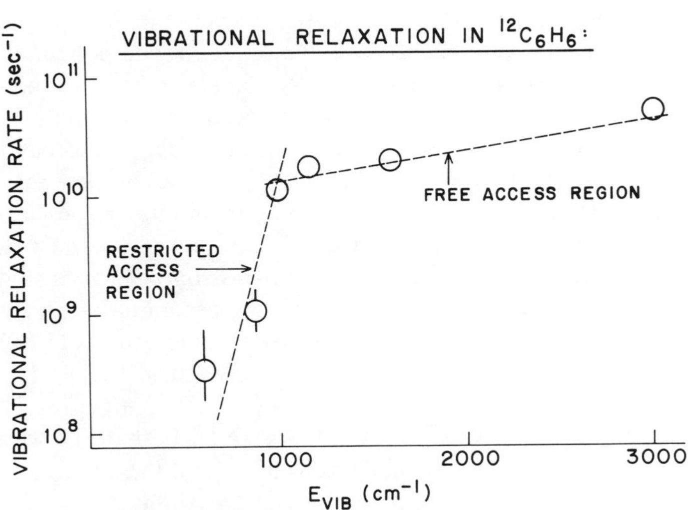

Fig. 8. Vibrational relaxation rates of some Raman-active modes in benzene. The dashed line is intended to draw the eye to the two regions of behavior of rate versus total vibrational energy.

It will be interesting to discover whether the trend of Fig. 8 is upheld by studies of the infrared active modes of benzene and whether these basic ideas carry over to other systems. The energy at which the anharmonic effects become dominant is expected to vary from system to system. For example, in larger molecules where there are many more lower-frequency modes, the onset of the free-access region may be at quite low total energy. The lowest-frequency modes may then be too close to the lattice modes for there to be a significant restricted region.

Molecular crystal vibrational transitions can be extremely sharp and, as a result of the exchange of vibrational energy, can display multiplets (factor group component states) of closely spaced lines. The study of the quantum beats in time-resolved CARS [see Eq. (69b)] was shown by Velsko et $al.$ (1983) to be an effective nonlinear method of ultrahigh resolution spectroscopy required to determine the position and widths of these lines. These experiments were carried out with ~ 5 ps pulses, but current femtosecond technology should enable the observation of the beats between and optical pumping of the molecular vibrational states.

As dilute impurities are added (or for neat crystals with natural abundances of isotopic impurities), line-broadening mechanisms due to energy trapping or scattering by impurities become operative. Simple theoretical models indicate that dephasing due to scattering is a slow process, except possibly in the resonant regime, where the impurity levels lie close to the host band. The lineshape for the dilute impurity takes on a form typical of impurity spectra: if the guest level is sufficiently far from nearby host bands, the lineshape will have a strong inhomogeneous component, which reflects the distribution of site energies due to crystal strain fields. The homogeneous component of the line reflects population decay, both to the surrounding host material and to lower modes of the guest molecules. If there is no spectral diffusion, the lineshape will be a convolution of the inhomogeneous and homogeneous bands.

At high doping levels ($\sim 50\%$), a number of processes become interwoven: impurity scattering may make an appreciable contribution to dephasing. The motional narrowing effect is reduced by the dilution of the host structure, allowing strains to contribute an inhomogeneous character to the linewidth. The population decay also changes character, becoming an "incoherent" process. All of these effects are not necessarily separable and additive.

As an example, Fig. 9 shows the CARS lineshape of the ν_1 A_g mode in crystalline benzene for a neat C_6H_6 crystal, for a 50% mixed C_6H_6/C_6D_6 crystal, and for dilute (3%) C_6H_6 in a C_6D_6 host. Note both the nonmonotonic dependence of overall linewidth on concentration and the qualitative change in the lineshapes. One expects different modes to exhibit different behavior, depending on where the $k = 0$ state lies within the band and on how close it lies to the levels of the isotopic diluent. Dlott and co-workers (Chronister and

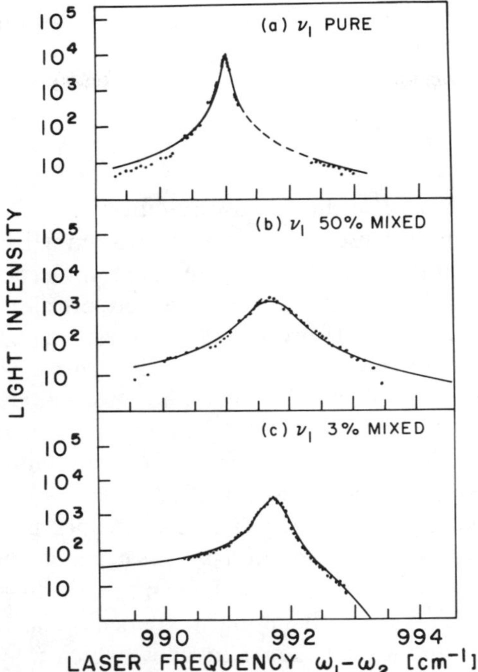

Fig. 9. Dependence of the benzene ν_1 A_g line width on concentration of perproto- in perdeuterobenzene. (a) Isotopically pure crystal with fit to a Lorentzian line shape, $\Gamma = 0.043$ cm^{-1}; the fits in the mixed-crystal spectra (b) and (c) are convolutions of gaussian and Lorentzian line shapes with parameters $\Gamma = 0.13$ cm^{-1}, $\sigma = 0.25$ cm^{-1}, and $\Gamma = 0.073$ cm^{-1}, $\sigma = 0.133$ cm^{-1} for the 50% and 3% crystals, respectively.

Dlott, 1983; Schosser and Dlott, 1984) have studied the concentration dependence of vibrational coherence decay times in mixtures of proto- and perdeuteronaphthalene, and have observed several distinct behaviors for different modes. It is clear that to interpret these dependences it is necessary to know in detail the vibrational levels of the host and guest molecules and their band structures, as well as the intrinsic strain-induced site energy distribution and its dependence on impurity concentration.

Vibrational energy relaxation and energy transfer to impurities are the most "chemically" interesting processes, since they relate to broader questions about vibrational relaxation in molecular systems. There are now a few cases where coherence decay can be unambiguously assigned to these mechanisms. In this respect an important experimental challenge is to make systematic studies of vibrational relaxation in crystals using direct T_1 measurements. Of course, from the point of view of exciton transport theory, dephasing processes are of interest in themselves, and also because of their potential for affecting the dynamics of population decay processes, especially trapping (Velsko and Hochstrasser, 1985a,b).

D. Resonances in $\chi^{(2)}$

The second-order electric dipole susceptibility was rarely used in spectroscopic applications because it vanishes in centrosymmetric media. Benzene, naphthalene, and anthracene are, as are most molecular crystals, centrosymmetric, and two-photon transitions are therefore allowed only between levels of the same parity. However, in all these materials two-photon transitions were noticed to occur from the totally symmetric ground state to levels known to be of u symmetry from one-photon spectra (Hochstrasser and Sung, 1977b; Hochstrasser et al., 1979). In addition, a resonance-enhanced second-harmonic generation was observed, corresponding again to $g \rightarrow u$ transitions (Hochstrasser and Meredith, 1977, 1978, 1979; Stevenson et al., 1981; Stevenson and Small, 1983). The anisotropic properties of these nonlinear effects suggested a two-photon process in which one photon couples through the electric and the other through the magnetic dipole interaction. In these experiments, fusion of the initially created polaritons generates a polariton at twice the incident laser frequency, which may either result in incoherent light (hyper-Raman scattering) or may survive as a coherent beam at 2ω. The measurements for a crystal of naphthalene are shown in Fig. 10. The sharp angular dependence of the peak of the signal maps the dispersion of the polariton: the generation of the second-harmonic beam is governed by the phase-matching conditions, and as 2ω resonates with a one-photon allowed transition, the index of refraction at this frequency varies rapidly not only as a function of the frequency (polariton dispersion) but also, in these biaxial crystals, as a function of the direction of the wave vector. Similar experiments have also been performed in the noncentric crystal phenanthrene (Johnson and Small, 1982a,b); in this system, transitions to the first electronic state are both one-photon and two-photon allowed and can thus be observed as resonances in $\chi^{(2)}$ and $\chi^{(3)}$. The quantitative analysis of the experimental data is rendered complex in noncentric systems because of interactions between incident and generated fields. In molecular crystals the problem of this cascading in higher-order processes has been critically analyzed by Meredith (1981, 1982). In centrosymmetric crystals these processes do not occur within the electric dipole approximation, and the forbidden second-order processes remain weak: in this case the breaking of the inversion symmetry by defects or by the surface may become important in contributing significantly to the generated signals.

At roughened metal surfaces, optical nonlinearities are strongly enhanced by the local field. This effect is thought to make significant contributions to

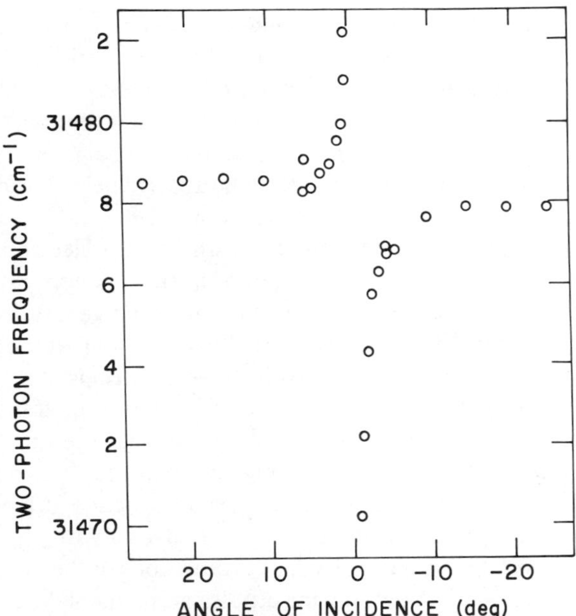

Fig. 10.   Variations of maxima of the second-harmonic signal as a function of laser-beam incidence angle in single crystals of naphthalene at 4.2 K. The angle is measured relative to the normal of the *ab* plane and the crystal is rotated about *b*; the polarization of the incident beam is parallel to *b*.

surface-enhanced Raman scattering (SERS). The contribution of adsorbed molecules to second harmonic generation also becomes very large such that molecular monolayers can easily be detected, as was shown by Shen and co-workers (Chen *et al.*, 1981; Heinz *et al.*, 1981, 1982, 1983). Such contributions become large even for centrosymmetric molecules: the second-order polarizability induced in the local field may reach values as large as those observed for polar molecules.

E. Fully Resonant $\chi^{(2)}$

In the presence of a dc electric field, $\chi^{(2)}$ phenomena such as sum- and difference-frequency generation will occur in all media. Although the observed effects are strictly third-order in the applied fields, two optical and one dc, the coherent light generation pathways and the dynamical parts of the susceptibility determining the spectral shapes are characteristic of $\chi^{(2)}$ processes. Using this approach, spectroscopic applications of both sum- and difference-frequency generation under fully resonant conditions have been explored (Dick and Hochstrasser, 1983d, 1984a). The system used to demonstrate these effects was a mixed crystal of a polar guest (azulene) doped substitutionally into a centrosymmetric host (naphthalene). The average dipole moment of the mixed crystal is zero, but in the presence of a dc electric field the random translational lattice of guest molecules is transformed into two interpenetrating but distinguishable sublattices consisting of polar molecules whose dipoles project parallel and antiparallel to the applied field. The optical transitions of the guest molecules in these two sublattices can then be separately observed. If this system is now subject to an intense electromagnetic field having an arbitrary frequency, it will respond as if it were noncentrosymmetric. Conventional electric-field-induced second-harmonic generation will occur, for example, with the SHG radiation intensity depending on the square of the dc-field strength. However, a qualitatively different effect is observed when the oscillating field is nearly resonant with one of the guest transitions corresponding to just one of the polar sublattices. In this case, the field senses a material that is polar, and at sufficiently high dc fields the $\chi^{(2)}$ process, occuring as a result of the response of one sublattice, becomes nearly independent of the dc field strength. The experiments were carried out with the optical field chosen to be resonant with spectrally sharp transitions of the S_1-S_0 and S_2-S_0 transitions of azulene. The permanent dipole moments of S_0, S_1, and S_2 are known to be sufficiently different that relatively small dc fields cause readily observable pseudo-Stark splittings of the spectral lines and effectively separate the two sublattices. The resonant contribution to $\chi^{(2)}$ of each sublattice separately are given by

$$\chi^{(2)}_{\text{SUM}} = \frac{\mu^{(1)}_{01}\mu^{(2)}_{12}\mu^{(3)}_{20}}{(\omega_{10}-\omega_1+i\Gamma_{01})[\omega_{20}-(\omega_1+\omega_2)+i\Gamma_{02}]} \tag{70}$$

$$\chi^{(2)}_{\text{DIF}} = \frac{\mu^{(2)}_{20}\mu^{(1)}_{01}\mu^{(3)}_{12}}{(\omega_{10}-\omega_1+i\Gamma_{01})(\omega_{20}-\omega_2-i\Gamma_{02})}\left\{1+\frac{i(\Gamma_{12}-\Gamma_{01}-\Gamma_{02})}{[\omega_{21}-(\omega_2-\omega_1)-i\Gamma_{12}]}\right\} \tag{71}$$

Resonances should therefore occur at ω_{10} and ω_{20} in both sum and difference frequency generation, and in addition a resonance at ω_{12} is predicted in the presence of pure dephasing (DICE resonance; see Section IV,G). In zero field, the contributions of the two sublattices have equal absolute values but are of opposite sign and cancel. In a dc field the shift of the transition frequencies ω_{ij} is opposite for the two sublattices, resulting in a net nonzero value of $\chi^{(2)}$, which can be obtained from Eqs. (70) and (71) as

$$\chi = \chi(\omega_{ij} - \Delta\boldsymbol{\mu}_{ij}\cdot\mathbf{F}) - \chi(\omega_{ij} + \Delta\boldsymbol{\mu}_{ij}\cdot\mathbf{F}) \tag{72}$$

where $\Delta\boldsymbol{\mu}_{ij}$ is the difference of permanent dipole moment of azulene in states i and j, and F the electric field strength along $\Delta\boldsymbol{\mu}$. Since the naphthalene host crystal is centrosymmetric, the nonresonant field induced value of $\chi^{(2)}$ is very small and can be neglected. This predicted behavior of $\chi^{(2)}$ is manifested in the experiment. Figure 11 shows examples of the field-induced sum- and difference-frequency generation signal; all predicted resonances except the DICE resonance are observed in these experiments, and the change of shape as a function of field strength can be fitted by formulas for $\chi^{(2)}$ as obtained from Eqs. (70)–(72). The absence of the DICE resonance was rationalized by numerical calculations of the spectra using parameters derived from linear spectroscopic data, which predict it to be more than three orders of magnitude weaker than the main resonance and to be thus unobservable under the experimental conditions. The ω_{20} and ω_{10} resonances appear very similar in the sum- and difference-frequency spectra, although complementary line-narrowing properties are predicted for both spectra. This is because the purely homogeneous contribution to the nonlinear linewidth dominates and any

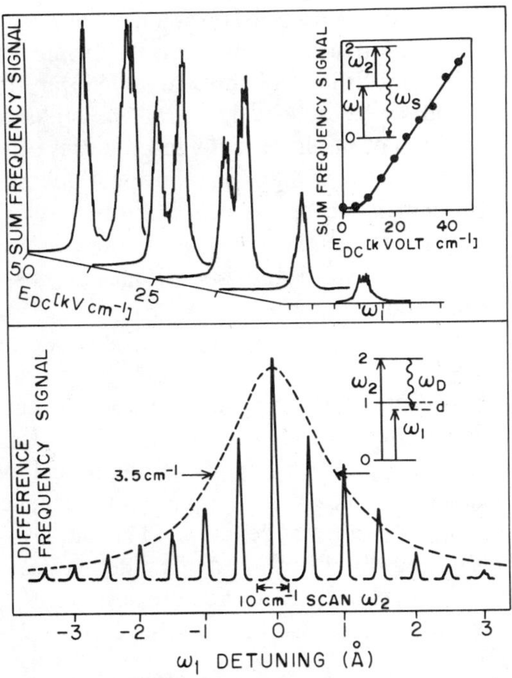

Fig. 11. Resonant sum- and difference-frequency generation of mixed crystals of azulene in naphthalene in an applied electric field. Top: Sum-frequency spectra for increasing dc field strength in the c' direction. Insert: The peak intensity of a sum-frequency signal as a function of dc field strength for low fields. Bottom: Difference-frequency generation in a dc field of 50 kV/cm along c' for various detunings of ω_1. The broken line is a fitted Lorentzian that maps the line shape of the $S_0 \rightarrow S_1$ transition.

correlations within the inhomogeneous contributions have only a slight effect in this case. In systems where the inhomogeneous width dominates, these experiments can be used for line narrowing and to study inhomogeneous correlations, as described in Section III.G and also discussed in the next section.

F. Four-Level Systems and Fully Resonant $\chi^{(3)}$

In order that a general four-wave mixing process be fully resonant, the system must have four levels, and each of these levels must be connected to two other levels by a one-photon allowed transition (see Fig. 3). This condition is very naturally met in molecules when the fundamental frequencies are in the range of electronic or vibronic transitions and when their differences equal vibrational or rotational frequencies. With only two different incident laser beams, just two resonance conditions can, in general, be fullfilled simultaneously, but in molecular systems a third resonance is also met approximately in as far as vibrational frequencies in different electronic states are usually similar (see 3–6 in Fig. 3). With use of three tunable input frequencies, exact resonance can of course be achieved for all possible transitions. The susceptibility of a medium of noninteracting molecules is proportional to the number density, and the measured signals are proportional to $|\chi|^2$. For a fully resonant four-level system, however, the susceptibility may become very large and not only allow the study of extremely dilute samples ($\sim 10^{-2}$ ppm) but also make it possible to address and study specific molecular species in selected states and/or environments.

The first fully resonant four-wave mixing studies were done in the dilute mixed crystal system of pentacene dispersed in benzoic acid (DeCola *et al.*, 1980a). The host, benzoic acid, forms high-quality crystals that are transparent not only to the visible frequencies ω_1 and ω_2 but also to $2\omega_1$ and $2\omega_2$. This system is expected to correspond closely to the theoretical model that treats two monochromatic waves at ω_1 and ω_2 coupled to a four-level system having resonances near $\omega_2, \omega_1, \omega_1 - \omega_2$, and $2\omega_1 - \omega_2$. The four levels correspond to the zeropoint levels (0 and 0') of ground and excited states and any pair of vibrational levels v and u', chosen from each of these states, for which $\mu_{vu'}$ is nonzero. The case where $v = u$, so that the same mode is involved in each state, was studied in detail.

Figure 12 shows experiments where ω_1 is fixed at the 0–0 transition, ω_2 tuned in the region of the fluorescence lines, and the generated beam at $2\omega_1 - \omega_2$ is monitored. The resonantly enhanced signals can be many times stronger than those from the host material, notwithstanding the fact that the observed light intensity varies with the square of the concentration. For this

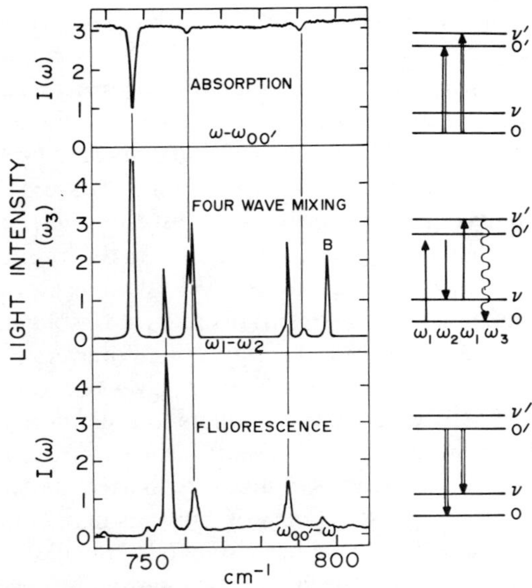

Fig. 12. Absorption, fluorescence, and CARS four-wave mixing spectra of pentacene in a benzoic acid crystal at 1.6 K. The four-wave mixing spectrum was measured on a crystal of 5×10^{-7} mol/mol concentration, and ω_1 was in exact resonance with the 0–0 transition of pentacene. The band B is a benzoic acid CARS resonance and can be used for calibration of the signal since it has a conventional off-resonance Raman cross section comparable with the 992 cm^{-1} mode of benzene.

case the enhancement factor was found to be greater than 10^{15}! The four-wave mixing signal is seen to contain information from both absorption and emission spectroscopy. Actually, the resonant part of the homogeneous response in this case takes the form

$$\sum_{u,v} \frac{(1/\Gamma_{00'})\mu_{00'}\mu_{0'v}\mu_{vu'}\mu_{u'0}}{[i(\omega_{u'} - \Delta) - \Gamma_{u'0}][i(\omega_v - \Delta) - \Gamma_{v0}]} \tag{73}$$

where ω_v and $\omega_{u'}$ are ground- and excited-state vibrational frequencies corresponding to modes v and u and $\Delta = \omega_1 - \omega_2$. The signal is the square of this function with peaks at $\Delta = \omega_v$ and $\Delta = \omega_{u'}$ for all modes u and v for which transition moments in the numerator exist. For many molecules, however, the dominant contribution comes from $v = u$. This results from the fact that $\mu_{vu'}$ is often maximum when $u = v$, especially in large molecule spectra. In addition, $\omega_v - \omega_{v'}$ is frequently small. Indeed, intense signals are obtained from modes that have nearly the same frequency in the ground and excited states since then both factors in the denominator of Eq. (73) become small at about the same value of Δ. An example, shown in Fig. 12, is the mode at $\omega_v = 762$, $\omega_{v'} = 761$

cm^{-1}, which shows only a weak absorption line but a relatively strong doublet in the coherent experiment. The states u' and v must correspond to the same chemical component of the system, otherwise $\mu_{vu'} = 0$. Thus this technique can be used to identify the occurence of different components in a mixture, different sites such as in mixed crystals, and matrix-isolated species or different aggregates of the same species. An example of the last case is the identification of the levels of dimers of pentacene in a host crystal of p-terphenyl. These pairs are observed at higher doping levels, e.g., 10^{-4} to 10^{-3}, and correspond to situations where two neighboring p-terphenyl host molecules are replaced by a pair of pentacene guests. The ratio of pairs to isolated monomers is proportional to the doping level and remains therefore very small under usual conditions. Each excited state of the monomer gives rise to three levels of the dimer, two singly excited levels (plus and minus combinations of the molecular excitations) and one doubly excited level. Linear spectroscopic methods do not expose the dimer transitions, which are buried under the monomer absorption. Using the high selectivity of fully resonant four-wave mixing, Levinsky and Wiersma (1982) have been able to locate all these levels for one dimer. Of particular interest is the fact that the resonance associated with the doubly excited level is fairly narrow, corresponding to a lifetime of at least 2 psec, for the relaxation of this state by excitation fusion.

G. DICE Effects

There are three fully resonant time orderings that contribute to the generation of steady-state Stokes radiation in a four-level system, as indicated in Fig. 13. The second diagram differs from the first in that the initial ω_1 field is interchanged in time with the ω_2 field. Both these processes involve introducing coherence in a level pair $v'0'$ (the excited-state Raman transition). Generally, when coherence is introduced into a pair of levels, the response function will exhibit a resonance at the transition frequency. While each of the two terms in Fig. 13 resonate on $\omega_{v'} = \omega_2 - \omega_1$, it turns out that the sum of these terms need not display this resonance. In fact, this excited-state Raman resonance is predicted to be absent if the quantity $\Gamma = \Gamma_{00'} + \Gamma_{0v'} - \Gamma_{0v'}$ is exactly zero. This is the case when the pure dephasing (elastic) parts of the coherence decay vanish. The response as a function of $\Delta = \omega_2 - \omega_1$ is given approximately by

$$(\omega_{v'} + d - \Delta + i\Gamma_{0v'})^{-1}[(\omega_v - \Delta + i\Gamma_v)^{-1}$$

$$+ \Gamma(\omega_{v'} - \Delta + i\Gamma_{v'})^{-1}(\omega_v - d - \Delta - i\Gamma_{0'v})^{-1}] \tag{74}$$

The absolute square of this function yields resonances at $\omega_{v'} + d$ and $\omega_v - d$, both of which have electronic transition widths, and resonances at ω_v and $\omega_{v'}$

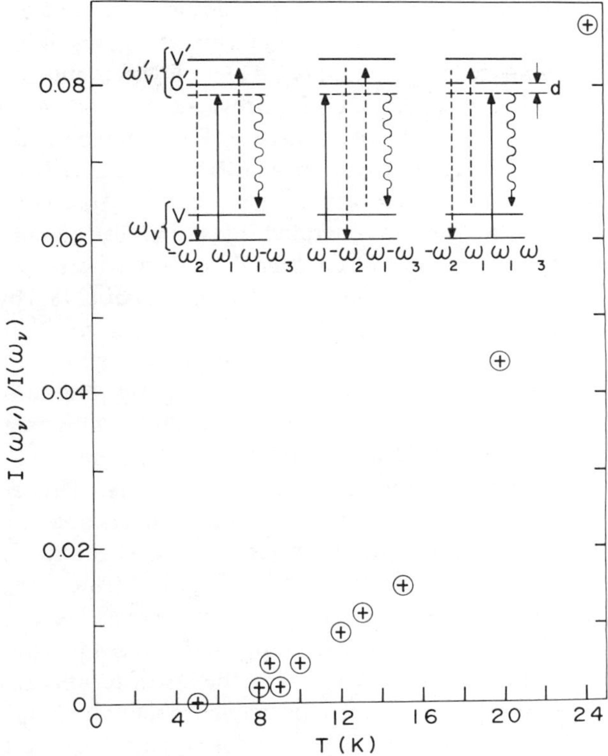

Fig. 13. Ratio of the signal intensity at $\Delta\omega = \omega_{v'}$ (747 cm^{-1}) and $\Delta\omega = \omega_v$ (755 cm^{-1}) in coherent Stokes Raman spectra as a function of temperature for pentacene in a benzoic acid crystal. The relevant diagrams for both resonances are inserted and are discussed in the text.

having the Raman widths of each of the electronic states. This interesting prediction implies that in the condensed phase at near the absolute zero of temperature where the pure dephasing might be small, one may expect little or no enhancement of the Stokes signal from the condition $\omega_2 - \omega_1 = \omega_{v'}$. On the other hand, at finite temperatures the excited-state Raman process is predicted to appear. This effect is termed dephasing induced coherent emission (DICE) (Andrews and Hochstrasser, 1981a,b). From the standpoint of spectroscopy it has important consequences, since inter-excited-state transitions can be studied without first *populating* the excited states. The populations are introduced by pure dephasing, or collisional redistribution in gases, in the same process that converts Raman scattering into fluorescence emission. The predicted temperature effects bring forth a method to measure directly the pure dephasing of transitions. Other coherent techniques such as photon echoes and hole-burning spectroscopy measure only the total dephasing rate.

The results for Stokes generation using pentacene in benzoic acid are given in Fig. 13. The DICE effect growth of intensity at the resonance condition $\omega_2 - \omega_1 = \omega_{v'}$ compared with that at $\omega_2 - \omega_1 = \omega_v$ is attributed to the onset of pure dephasing in the system. DICE processes are, in fact, accounted for in the usual form of the resonant susceptibility (Bloembergen *et al.*, 1978) and were recently seen also in atomic vapors as a PIER-4 effect (Prior *et al.*, 1981).

A question inherent in these experiments concerns the contribution of excited-state populations to the generated signal. In fact, the same dephasing process that brings about the DICE signal converts the polarization into population. This question was investigated by Bozio *et al.* (1983) by measuring CARS and CSRS spectra as a function of detuning d. At 1.6 K and 4.2 K and for small values of d (<5 cm^{-1}), the spectra show resonances at $|\omega_1 - \omega_2| = \omega_{v'}$ independent of d, demonstrating the buildup of excited-state population during the laser pulse. In the DICE experiment discussed above, d was large enough (16.8 cm^{-1}) and the contribution of the excited-state populations to the observed signals were estimated to be negligible. However, line broadening at higher temperatures makes population buildup possible for larger detuning, but comparisons of CARS and CSRS studies allow in this case the identification of the DICE effect (Andrews *et al.*, 1981).

The nonlinear responses of these four-level systems were also studied directly in time-domain experiments (Duppen *et al.*, 1983). The two Stokes diagrams 5 and 6 in Fig. 3, which give rise to the DICE resonance in quasi-cw (continuous wave) experiments can, in principle, be separated in time-domain studies with light pulses that are short compared with the system relaxation times. The time behavior corresponding to these two diagrams is

$$\exp[-(i\omega_{ac} + \Gamma_{ac})(t_1 - t_2) - (i\omega_{dc} + \Gamma_{dc})(t_2 - t_3) - (i\omega_{bc} + \Gamma_{bc})t_3] \quad (75)$$

$$\exp[-(i\omega_{da} + \Gamma_{da})(t_1 - t_2) - (i\omega_{dc} + \Gamma_{dc})(t_2 - t_3) - (i\omega_{bc} + \Gamma_{bc})t_3] \quad (76)$$

where in Eq. (75) a pulse at ω_2 arrives at $t_1 = t - \tau_1 = 0$ resonant or near resonant with the $a \to c$ transition, while in Eq. (76) ω_1 arrives at $t = \tau_1$ near resonant with the $a \to d$ transition. For the case where the ω_1 and ω_2 pulses arrive together, $\tau_1 = \tau_2$, a probe pulse at ω_2 generates Stokes light whose intensity variation with pulse delay, $\tau = \tau_3 - \tau_2 = t_2 - t_3$, measures the excited-state dynamics Γ_{cd}. The response function for both diagrams becomes in this case the same and is equal to

$$R(t) = -i\mu_{ac}\mu_{da}\mu_{bd}\mu_{cb} \exp[-(i\omega_{bc} + \Gamma_{bc})(t - \tau)]\exp[-(i\omega_{dc} + \Gamma_{dc})\tau] \quad (77)$$

The integrated intensity of the signal generated at ω_{bc} is equal to

$$I(\tau) = c \int_\tau^\infty dt_3 |R(t_3)|^2 = I_0 e^{-2\Gamma_{dc}\tau} \quad (78)$$

with $I_0 = c/2\Gamma_{bc}$ being the intensity of the signal for $\tau = 0$. Notice that this experiment does not depend on the existence of pure dephasing, so that the DICE effect is seen to be a result of averaging over times long compared with the relaxation dynamics.

One interesting aspect of these fully resonant processes is their potential for exploring inhomogeneous distributions (see Section III.G). For the $755/747$ cm^{-1} mode of bentacene in benzoic acid at 1.6 K, it was found that the width of the ground state Raman resonance is about two times smaller in CARS than in CSRS spectra (Bozio *et al.*, 1983). This is a clear indication of the fact that the inhomogeneous distributions of the ground-state Raman and the electronic transitions are anticorrelated: that is, molecules absorbing at the higher frequency side of the $0-0'$ band center contribute to the lower-energy portion of the inhomogeneous ground-state vibrational transition at 755 cm^{-1}.

H. Grating Experiments

As the frequency difference between the incident laser beams becomes smaller, lower-frequency molecular modes and optical and eventually acoustical phonons are probed. In fact, these experiments probe in general phenomena occuring on the timescale of $1/\Delta\omega$. These coherent Rayleigh mixing experiments have been used to probe dynamical processes in liquids (Yajima *et al.*, 1976, 1978; Yajima and Souma, 1978; Heilweil *et al.*, 1980; Souma *et al.*, 1982). The way in which these experiments probe the dynamics can be envisioned by considering the spatial distribution of light intensity set up in the sample by two intersecting incident beams: for two beams of equal frequency ω and field strength E crossing in the sample at an angle of 2θ, the intensity in the sample is spatially periodic

$$|E_1 + E_1'|^2 = 4E^2 \cos^2(2\pi y \sin\theta/\lambda) \tag{79}$$

corresponding to a stationary diffraction grating along the direction y that is parallel to $\mathbf{k}_1 - \mathbf{k}_1'$. Another beam ω_2 will diffract from this grating with the overall process leading to the generation of a beam at ω_2 in the direction $\mathbf{k}_1 - \mathbf{k}_1' + \mathbf{k}_2$. It is easy to see that this corresponds to the Bragg diffraction condition for a grating having the character of Eq. (79). When the frequencies of the two beams creating the grating differ by an amount $\Delta\omega = \omega_1 - \omega_2$, the spatial intensity distribution is no longer stationary and the diffracted beam will be Doppler shifted by an amount $\Delta\omega$. The diffraction of ω_1 gives rise to light emitted at frequency $2\omega_1 - \omega_2$. The amplitude of the grating and therefore the intensity of the diffracted beam is determined by the variation of the refractive index of the medium that is induced by the incident E_1 fields. Different contributions to this variation stem from thermal, population, and coherence transfer processes, and these contributions build up and decay with different characteristic time constants. In a moving grating, therefore, only contributions with characteristic time constants faster than $1/\Delta\omega$ will show up: otherwise they will be washed out. Thus, the dynamics of the medium can be probed by varying the frequency difference of the two beams creating the grating. For example, with incident visible beams separated by 1 cm^{-1}, processes occuring on the time scale of 5 psec are singled out.

The time-domain analogue of these moving-grating experiments are the so-called transient-grating experiments. In these experiments, two time-coincident laser pulses of the same frequency are crossed in the sample and produce a grating, which is probed by a delayed third pulse. The grating represents a spatially periodic modulation of the index of refraction, which has contributions from the population of excited electronic or vibrational states or phonons (or more generally of other species produced in the excitation process such as electron–hole pairs, for example, in semiconductors) and due to modulations of the density of the material produced by periodic heating or directly by electrostriction. The buildup of the grating amplitude occurs over a finite time and its observation permits study of the kinetic processes contributing to it. The probe pulse need not be of the same frequency as the excitation pulses and can be chosen to be resonant with a different transition and thereby to enhance specific contributions to the grating. In molecular crystals, transient-grating experiments have been used to study excited-state lifetimes, excited state absorptions, and exciton transport properties (Eichler, 1977; Fayer, 1982, 1983; Rose *et al.*, 1984).

Transient gratings have also been used to stimulate ultrasonic waves and to measure their speed and attenuation (Nelson *et al.*, 1982); these experiments correspond to Brillouin scattering in the time domain and are particularly useful when the attenuation becomes very large and difficult to measure in the frequency domain. For example, this method was employed to follow soft modes near a phase transition (Robinson *et al.*, 1984).

In transparent media the coupling to the light field occurs by electrostriction, but the generation of coherent acoustic phonons in optically absorbing media is more readily achieved by impulsive heating. These techniques to generate and monitor ultrasonic waves are superior to conventional methods, especially when the properties of the material make it difficult to establish a good mechanical contact with a transducer, or when the acoustic attenuation becomes so high that the wave is damped over only a few cycles. These situations arise frequently when a material undergoes interesting structural transformations: in polymer materials, for example, the glass transition as a function of temperature or the transformation from monomer to polymer during the polymerization may be studied using these methods, as is illustrated in Fig. 14 (Blanchard *et al.*, 1985).

Transient-grating experiments are intimately related to stimulated photon echo experiments: they correspond to the limit in which the time difference between the first two pulses in the echo experiment goes to zero. The situation in which the state of the grating is probed by a different color corresponds to two-color photon echo experiments (Duppen *et al.*, 1984a,b). Both experiments probe the same physical processes, namely the evolution of the population of the levels connected by the light fields.

I. Molecular Reorientation

The populations of excited states generated by nonlinear optical interactions are usually anisotropic. This anisotropy is manifested as a dichroism. Nonlinear signals that depend on such dichroism have response functions that

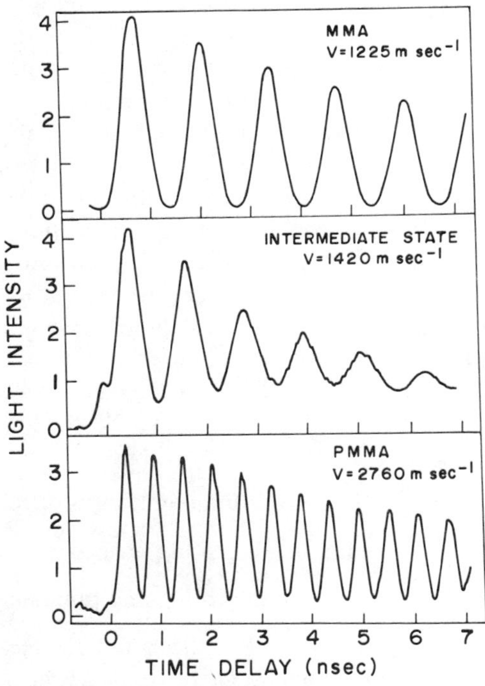

Fig. 14. Time evolution of a standing ultrasonic wave generated by impulsive heating by two time-coincident crossed laser pulses in methylmethacrylate during polymerization. While the speed of sound increases monotonically during this process, the acoustic attenuation becomes very large and passes through a maximum at an intermediate state of polymerization.

decay on the reduced time scale of T_1 relaxation and the orientational correlation time of the transition dipoles that generate the resonances. One example of such a signal is found in the polarization spectroscopy of molecules in solutions (Shank and Ippen, 1975; Reiser and Laubereau, 1982; Cross *et al.*, 1983; Myers and Hochstrasser, 1986). Motional properties of systems are usefully described using the conventional susceptibilities in cases where the motion is slow compared with the electromagnetic field impulses and T_2 decay times. In that case the response function of the system is developed by iteration in the usual way, except that the transition dipole factor is taken to be time-dependent but slowly varying (Myers and Hochstrasser, 1986).

For a system of four levels, a and b coupled by the excitation fields 1 and 2 at time 0, c and d coupled by the probing field, 3, at time τ, the usual density matrix expansion yields a slowly varying portion of the response to linear polarized fields having the form

$$\langle \mu_{ab}(0) \cdot \hat{e}_1 \mu_{ba}(0) \cdot \hat{e}_2 \mu_{cd}(\tau) \cdot \hat{e}_3 \mu_{dc}(\tau) \cdot \hat{e}_4 \rangle = |\mu_{ab}\mu_{cd}|^2 (A + Be^{-\tau/\tau_R}) \quad (80)$$

where $\hat{e}_4$ is the polarization of the generated wave. This response refers to all experiments involving a field product $\mathscr{E}_1 \mathscr{E}_1 ^* \mathscr{E}_2$, including polarization spectroscopy, transient gratings, and crossed gratings. Conventional Bragg diffraction in which a transient grating formed by excitation pulses of the same polarization ($\hat{e}_1 \cdot \hat{e}_2 = 1$) is probed with parallel ($\hat{e}_3 \cdot \hat{e}_2 = 1$) and perpendicular ($\hat{e}_3 \cdot \hat{e}_2 = 0$) polarizations has the advantage of yielding both the time dependence and the magnitude of the induced anisotropy. However, the accompanying acoustic grating interferes with the polarization-sensitive excited-state grating. In the crossed-grating configuration, the acoustic signal is eliminated by the use of perpendicularly polarized excitation pulses $\hat{e}_1 \cdot \hat{e}_2 = 0$ Polarization spectroscopy ($\hat{e}_1 \cdot \hat{e}_2 = 0$; $\hat{e}_4 \cdot \hat{e}_3 = 0$) gives the same susceptibility and anisotropy dynamics as the crossed grating but is more sensitive to interference from background birefringence because $\mathbf{k}_4 = \mathbf{k}_3$.

The orientation average of Eq. (80) is readily evaluated for the case of isotropic diffusion having relaxation time τ, to yield the following values for the constants A and B:

$A = \tfrac{1}{3}$;　　　$B = 2r_0/3$　　　　for conventional grating probed parallel

$A = \tfrac{1}{3}$;　　　$B = -r_0/3$　　　　for crossed grating probed perpendicular

$A = 0$;　　　$B = r_0/2$　　　　for crossed grating and polarization spectroscopy

where $r_0 = \tfrac{2}{5} \langle P_2[\mu_{ab}(0) \cdot \mu_{cd}(\tau)] \rangle$ and P_2 is the second Legendre polynomial. It is easy to deduce the relative signal intensities (at time zero) for each of these techniques. The results are given in Table I. These techniques are each useful in determining rotational relaxation dynamics in molecular condensed phases. Furthermore, when the four waves all have the same frequency, the methods

TABLE I

Angular Averages and Relative Signal Strengths for Different Four-Wave Mixing Configurations

	Ordinary grating, parallel probe[a]	Ordinary grating, perpendicular probe[a]	Crossed grating	Polarization spectroscopy
$\hat{\mathbf{e}}_1\hat{\mathbf{e}}_2\hat{\mathbf{e}}_3\hat{\mathbf{e}}_4$	$zzzz$	$zzyy$	$zyzy^b$	$zz(z+y)(z-y)\frac{1}{2}$
$\langle \mu_i^{(1)}(0)\mu_i^{(2)}(0)\mu_j^{(3)}(T)\mu_j^{(4)}(T)\rangle^c$	$\frac{1}{3}(1+2r_0e^{-T/\tau_R})$	$\frac{1}{3}(1-r_0e^{-T/\tau_R})$	$\frac{1}{2}r_0e^{-T/\tau_R}$	$\frac{1}{2}r_0e^{-T/\tau_R}$
Signal at $T=0$ for $\hat{\boldsymbol{\mu}}_i\cdot\hat{\boldsymbol{\mu}}_j=1$	9	1	1	4
Signal at $T=0$ for $\hat{\boldsymbol{\mu}}_i\cdot\hat{\boldsymbol{\mu}}_j=0$	1	4	0.25	1

[a] The field polarization and angular averages for fluorescence with parallel and perpendicular detection are the same as for the ordinary grating. The parallel : perpendicular signal ratios for fluorescence are the square root of those for the grating.

[b] In the crossed grating, any choice of $\hat{\mathbf{e}}_3$ gives the same signal strength. In general, if $\hat{\mathbf{e}}_3 = \cos\theta\hat{\mathbf{z}} + \sin\theta\hat{\mathbf{y}}$, then $\hat{\mathbf{e}}_4 = \sin\theta\hat{\mathbf{z}} + \cos\theta\hat{\mathbf{y}}$.

[c] Assuming isotropic rotational diffusion with time constant τ_R; $r_0 = \frac{2}{5}\langle P_2[\hat{\boldsymbol{\mu}}_i\cdot\hat{\boldsymbol{\mu}}_j]\rangle$.

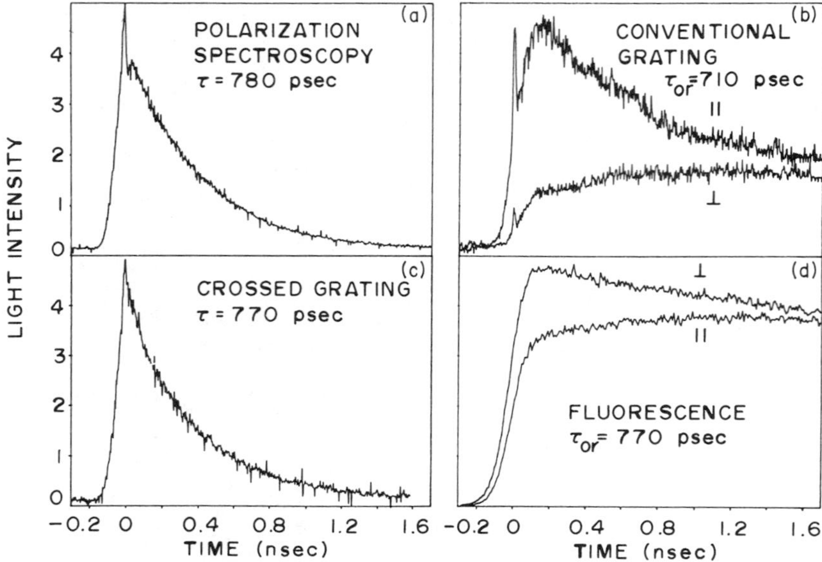

Fig. 15. Comparison of different techniques measuring the rotational relaxation dynamics of 9-aminoacridine in ethylene glycol. The excitation wavelength in all experiments is 266 nm. In (a)–(c), the coherence introduced in the sample is probed with a time-delayed pulse of the same frequency. For comparison in (d), the incoherent fluorescence of the sample detected at right angles at 460 nm is also shown. The difference of the two signals in (c) and (d) should measure the same dynamical process as each of the signals in (a) and (b).

provide a way of measuring T_2 for the pumped transition through studies of the "coherent spike" seen at zero delay time (see Fig. 15).

Figure 15 shows the experimental results for 9-aminoacridine using each of the grating configurations discussed above. The signal in each case corresponds to the square of the response function in Eq. (80). The deduced value of $\tau_R = 760 \pm 20$ psec obtained from each experiment corresponds in this case to the rotation of the molecule about an axis perpendicular to the molecular plane.

V. SUMMARY AND PROGNOSIS

This article was intended to provide an overview of nonlinear optical studies of molecular systems with particular emphasis on molecular condensed matter spectroscopic applications. We have shown that a wide range of material parameters can be determined by means of either time or frequency domain experiments that employ tunable lasers. The use of multiple laser fields allows the study of double or triple resonance effects and of dynamical processes involving molecular and excitonic excited states. The nonlinear optical theory provides the framework on which to understand all such multiresonant phenomena including induced emission, absorption, and spontaneous decay.

For the future, there are a number of obvious directions that resonant molecular optics is likely to take. First, there is the study of extremely rapid responses using femtosecond light pulses more closely approximating those of Eq. (7). Recently, the coherent Raman beating implicit in Eq. (69) and observed for benzene using picosecond pulses (Velsko *et al.*, 1983) was studied with femtosecond pulses by Nelson and coworkers (De Silvestri *et al.*, 1985). In addition it seems very interesting to proceed with the study of systems not close to equilibrium. This situation might involve materials that are heavily ionized or highly concentrated in excited states or excitons and systems in which a large fraction of certain atoms, for example protons, are displaced from their equilibrium configurations. Another area of great interest barely touched on in this article is surface molecular optics. It is apparent that the principles presented here can be used to make detailed studies of surface states and of dynamical processes involving adsorbed molecules. Finally, an important area is the development and characterization of new molecular materials having the required resonances and dynamics to permit qualitative improvements in, as well as the generation of, new types of optical materials.

ACKNOWLEDGMENTS

This research was supported by grants from NSF, NIH, AROD, and the LRSM Program at Pennsylvania. We are indebted to our colleagues in the Penn Molecular Optics group for their contributions and encouragement.

REFERENCES

Abram, I., and Hochstrasser, R. M. (1979). *In* "Light Scattering in Solids" (J. L. Birman, ed.), p. 447. Plenum, New York.

Abram, I., and Hochstrasser, R. M. (1980). *J. Chem. Phys.* **72,** 3617.

Allen, L., and Eberly, J. H. (1975). "Optical Resonance and Two-Level Atoms." Wiley, New York.

Andrews, J. R., and Hochstrasser, R. M. (1981a). *Chem. Phys. Lett* **82,** 381.

Andrews, J. R., and Hochstrasser, R. M. (1981b). *Chem. Phys. Lett.* **83,** 427.

Andrews, J. R., Hochstrasser, R. M., and Trommsdorff, H. P. (1981). *Chem. Phys.* **62,** 87.

Bechtel, I. H., and Smith, W. L. (1976). *Phys. Rev. B* **13,** 3515.

Blanchard, D., Casalegno R., Pierre, M., and Trommsdorff H. P. (1985). *J. Phys.* **46,** C7–517.

Bloembergen, N. (1965). "Nonlinear Optics." Benjamin, New York.

Bloembergen, N., Lotem, H., and Lynch, R. T., Jr. (1978) *Indian J. Pure Appl. Phys.* **16,** 151.

Bordé, C. J. (1976). *C. R. Hebd. Seances Acad. Sci., Ser. B* **282,** 341.

Bordé, J., and Bordé, C. J. (1978). *J. Mol. Spectrosc.* **78,** 353.

Bozio, R., DeCola, P. L., and Hochstrasser, R. M. (1983). *In* "Time-Resolved Vibrational Spectroscopy" (G. H. Atkinson, ed.), p. 335. Academic Press, New York.

Burns, M. J., Lin, W. K., and Zewail, A. H. (1983). *In* "Spectroscopy and Excitation Dynamics of Condensed Molecular Systems" (V. M., Agranovich, and R. M., Hochstrasser, eds.), p. 301. North-Holland Publ., Amsterdam.

Burris, J., McGee, T. J., and McIlrath, T. J. (1983). *Chem. Phys. Lett.* **101,** 588.

Butcher, P. N. (1965). "Nonlinear Optical Phenomena." Ohio State University Engineering Publications, Columbus.

Chen, C. K., Heinz, T. F., Richard, D., and Shen, Y. R. (1981). *Phys. Rev. Lett.* **46,** 1010.

Chronister, E. L., and Dlott, D. D. (1983). *J. Chem. Phys.* **79,** 5286.

Cross, A. J., Waldeck, D. H., and Fleming, G. R. (1983). *J. Chem. Phys.* **78,** 6455.

Davydov, A. S. (1971). "Theory of Molecular Excitons," Plenum, New York.

DeCola, P. L., Andrews, J. R., Hochstrasser, R. M., and Trommsdorff, H. P. (1980a). *J. Chem. Phys.* **73,** 4695.

DeCola, P. L., Hochstrasser, R. M., and Trommsdorff, H. P. (1980b). *Chem. Phys. Lett.* **72,** 1.

De Silvestri, S., Fujimoto, J. G., Ippen, E. P., Gamble, E. B., Jr., Williams, L. R., and Nelson, K. A. (1985). *Chem. Phys. Lett.* **116,** 146.

De Vries, H., and Wiersma, D. A. (1976). *Phys. Rev. Lett.* **36,** 91.

Dick, B., and Hochstrasser, R. M. (1983b). *J. Chem. Phys.* **78,** 3398.

Dick, B., and Hochstrasser, R. M. (1983b). *Chem. Phys.* **75,** 133.

Dick, B., and Hochstrasser, R. M. (1983c). *Chem. Phys. Lett.* **102,** 484.

Dick, B., and Hochstrasser, R. M. (1983d). *Phys. Rev. Lett.* **51,** 2221.

Dick, B., and Hochstrasser, R. M. (1984a). *Chem. Phys.* **91,** 1.

Dick, B., and Hochstrasser, R. M. (1984b). *J. Chem. Phys.* **81,** 2897.

Druet, S. A. J., Attal, B., Gustafson, T. K., and Taran, J. P. E. (1978). *Phys. Rev. A* **18,** 1529.

Duppen, K., Weitekamp, D. P., and Wiersma, D. A. (1983). *J. Chem. Phys.* **79,** 5835–5844.

Duppen, K., Weitekamp, D. P., and Wiersma, D. A. (1984a). *Chem. Phys. Lett.* **106,** 141.

Duppen, K., Weitekamp, D. P., and Wiersma, D. A. (1984b). *Chem. Phys. Lett.* **108,** 551.

Eichler, E. J. (1977). *Opt. Acta* **24,** 431.

Fayer, M. D. (1982). *Annu. Rev. Phys. Chem.* **33,** 631.

Fayer, M. D. (1983). *In* "Spectroscopy and Excitation Dynamics of Condensed Molecular Systems," (V. M. Agranovich and R. M. Hochstrasser, eds.), p. 185 North-Holland Publ., Amsterdam.

Fendt, A., Fischer, S. F., and Kaiser, W. (1981). *Chem. Phys.* **57,** 55.

Flytzanis, C. (1975). *In* "Quantum Electronics: A Treatise" (H. Rabin and C. L. Tang, eds.), Vol. 1A, Part A, p. 1. Academic Press, New York.

Heilweil, E. J., Hochstrasser, R. M., and Souma, H. (1980). *Opt. Commun.* **35,** 227.

Heinz, T. F., Chen, C. K., Richard, D., and Shen, Y. R. (1981). *Chem. Phys. Lett.* **83,** 180.

Heinz, T. F., Chen, C. K., Richard, D., and Shen, Y. R. (1982). *Phys. Rev. Lett.* **48,** 478.

Heinz, T. F., Tom, H. W. K., and Shen, Y. R. (1983). *Phys. Rev. A* **28,** 1883.

Hellwarth, R. W. (1977). *Prog. Quantum Electron* **5,** 1.

Hesp, B. H., and Wiersma, D. A. (1980). *Chem. Phys. Lett.* **75,** 423.

Hesselink, W. H., and Wiersma, D. A. (1983). *In* "Spectroscopy and Excitation Dynamics of Condensed Molecular Systems" (V. M. Agranovich and R. M. Hochstrasser, eds.), p. 249. North-Holland Publ., Amsterdam.

Ho, F., Tsay, W. S., Trout, J., and Hochstrasser, R. M. (1981). *Chem. Phys. Lett.* **83,** 5.

Ho, F., Tsay, W. S., Trout, J., Velsko, S., and Hochstrasser, R. M. (1983). *Chem. Phys. Lett.* **97,** 141.

Hochstrasser, R. M. (1973). *Acc. Chem. Res.* **6**, 263.
Hochstrasser, R. M. (1976). *Int. Rev. Sci.: Phys. Chem., Ser. 2* **3**, 1.
Hochstrasser, R. M., and McAlpine, R. D. (1966). *J. Chem. Phys.* **44**, 3325.
Hochstrasser, R. M., and Meredith, G. R. (1977). *J. Chem. Phys.* **67**, 1273.
Hochstrasser, R. M., and Meredith, G. R. (1978). *Pure Appl. Chem.* **50**, 759.
Hochstrasser, R. M., and Meredith, G. R. (1979). *J. Luminescence* **18/19**, 32.
Hochstrasser, R. M., and Sung, H. N. (1977a). *J. Chem. Phys.* **66**, 3265.
Hochstrasser, R. M., and Sung, H. N. (1977b). *J. Chem. Phys.* **66**, 3276.
Hochstrasser, R. M., Sung, H. N., and Wessel, J. E. (1973a). *J. Am. Chem. Soc.* **95**, 8179.
Hochstrasser, R. M., Sung, H. N., and Wessel, J. E. (1973b). *J. Chem. Phys.* **58**, 4694.
Hochstrasser, R. M., Wessel, J. E., and Sung, H. N. (1974). *J. Chem. Phys.* **60**, 317.
Hochstrasser, R. M., Klimcak, C. M., and Meredith, G. R. (1979). *J. Chem. Phys.* **70**, 870.
Hochstrasser, R. M., Meredith, G. R., and Trommsdorff, H. P. (1980). *J. Chem. Phys.* **73**, 1009.
Hudson, B. S., Kohler, B. E., and Schulten, K. (1982). *In* "Excited States" (E. C. Lim, ed.), Vol. 6, p. 1. Academic Press, New York.
Johnson, C. K., and Small, G. J. (1982a). *J. Chem. Phys.* **76**, 3837.
Johnson, C. K., and Small, G. J. (1982b). *Chem. Phys.* **64**, 83.
Klafter, J; and Jortner, J. (1977). *Chem. Phys. Lett.* **50**, 202.
Klafter, J., and Jortner, J. (1978). *J. Chem. Phys.* **68**, 1513.
Kubo, R. (1969). *Adv. Chem. Phys.* **15**, 101.
Kubo, R., and Tomito, K. (1954). *J. Phys. Soc. Jpn.* **9**, 888.
Levenson, M. D., and Bloembergen, N. (1974a). *J. Chem. Phys.* **60**, 1323.
Levenson, M. D., and Bloembergen, N. (1974b). *Phys. Rev. B* **10**, 4447.
Levinsky, H., and Wiersma, D. A. (1982). *Chem. Phys. Lett.* **92**, 24.
Liptay, W. (1974). *In* "Excited States" (E. C. Lim, ed.), Vol. 1, p. 129. Academic Press, New York.
Lotem, H., and Lynch, R. T., Jr. (1976), *Phys. Rev. Lett.* **37**, 334.
Lynch, R. T., Jr., and Lotem, H. (1977), *J. Chem. Phys.* **66**, 1905.
McAlpine, R. D. (1968). Ph.D. Dissertation, University of Pennsylvania, Philadelphia.
McClain, W. M., and Harris, R. A. (1977). *In* "Excited States" (E. C. Lim, ed.), Vol. 3. Academic Press, New York.
Meredith, G. R. (1981). *J. Chem. Phys.* **75**, 4317.
Meredith, G. R. (1982). *J. Chem. Phys.* **77**, 5863.
Moya, F., Druet, S. A. J., and Taran, J. P. E. (1975). *Opt. Commun.* **13**, 169.
Mulliken, R. S., and Person, (1969). "Molecular Complexes." Wiley, New York.
Myers, A. B., and Hochstrasser (1986). *IEEE J. Quantum Electron.* **QE22**, 1482.
Nelson, K. A., Casalegno, R., Miller, R. J. D., and Fayer, M. D. (1982). *J. Chem. Phys.* **77**, 1144.
Oudar, J.-L., and Shen, Y. R. (1980), *Phys. Rev. A* **22**, 1141.
Prior, Y., Bogdan, A. R., Dagenais, M., and Bloembergen, N. (1981). *Phys. Rev. Lett.* **46**, 111.
Rabi, I. I. (1937). *Phys. Rev.* **51**, 652.
Regnier, P., and Taran, J. P. E. (1973). *Appl. Phys. Lett.* **23**, 240.
Regnier, P., Moya, F., and Taran, J. P. E. (1974). *AIAA J.* **12**, 826.
Reiser D., and Laubereau, A. (1982). *Chem. Phys. Lett.* **92**, 297.
Righini, K., Fracassi, P. F., and Della Valle, K. G. (1983). *Chem. Phys. Lett.* **97**, 308.
Robinson, G. W. (1970). *Ann. Rev. Phys. Chem.* **21**, 429.
Robinson, M. M., Yan, Y.-X., Gamble, E. B., Jr., Williams, L. R., Meth, J. S., and Nelson, K. A. (1984). *Chem. Phys. Lett.* **112**, 491.
Rose, T. S., Righini, R., and Fayer, M. D. (1984). *Chem. Phys. Lett.* **106**, 13.
Schmalz, T. G., and Flygare, W. H. (1978). *In* "Laser and Coherence Spectroscopy" (J. I. Steinfeld, ed.), pp. 125–196. Plenum, New York.
Schosser, C. L., and Dlott, D. D. (1984). *J. Chem. Phys.* **80**, 1394.

Shank, C. V., and Ippen, E. P. (1975). *Appl. Phys. Lett.* **26,** 62.

Shen, Y. R. (1974). *Phys. Rev. B* **9,** 622.

Small, G. J. (1983). *In* "Spectroscopy and Excitation Dynamics of Condensed Molecular Systems" (V. M. Agranovich and R. M. Hochstrasser, eds.), p. 515. North-Holland Publ., Amsterdam.

Souma, H., Heilweil, E. J., and Hochstrasser, R. M. (1982). *J. Chem. Phys.* **76,** 5693.

Stevenson, S. H., and Small, G. J. (1983). *Chem. Phys. Lett.* **100,** 334.

Stevenson, S. H., Johnson, C. K., and Small, G. J. (1981). *J. Phys. Chem.* **85,** 2709.

Trout, T. J., Velsko, S., Bozio, R., DeCola, P. L., and Hochstrasser, R. M. (1984). *J. Chem. Phys.* **81,** 4746.

Velsko, S., and Hochstrasser, R. M. (1985a). *J. Phys. Chem.* **89,** 2240.

Velsko, S., and Hochstrasser, R. M. (1985b). *J. Chem. Phys.* **82,** 2180.

Velsko, S., Trout, J., and Hochstrasser, R. M. (1983). *J. Chem. Phys.* **79,** 2114.

Whiteman, J., McAlpine, R. D., and Hochstrasser, R. M. (1973). *J. Chem. Phys.* **58,** 5078.

Yajima, T., and Souma, H. (1978). *Phys. Rev. A* **17,** 309.

Yajima, T., Souma, H., and Ishida, Y. (1976). *Opt. Commun.* **18,** 150.

Yajima, T., Souma, H., and Ishida, Y. (1978). *Phys. Rev. A* **17,** 324.

Yee, S. Y., and Gustafson, T. K. (1978). *Phys. Rev. A* **18,** 1597.

Yee, S. Y., Gustafson, T. K., Druet, S. A. J., and Taran, J. P. E. (1977). *Opt. Commun.* **23,** 1.

Part IV

Concluding Comments

Chapter IV-1

Optical Properties and the Intermolecular Bond: By Way of Extension from Molecular to Supramolecular Materials

JEAN-MARIE LEHN

Collège de France, Paris; and Université Louis Pasteur, Strasbourg

The purpose of this brief contribution is to speculate about the potential interest for nonlinear optics of classes of compounds and types of processes to which, in this respect, little, if any, consideration has been given. It concerns coordination compounds, metal complexes, molecular acceptor–donor complexes, etc.—those chemical species that are held together by intermolecular bonding and form what may be termed supramolecular[1] assemblies.

At the outset, however, a cautionary note is warranted. The picture presented here will not be an organized, exhaustive account but will leave much to imagination. Rather it is intended to point out relationships, similarities, and possible meeting points in approach, materials, or goals between fields that may appear at first to have little in common or for which

215

these common interests have not been stressed. Since vast domains will be touched upon, the bibliography will be highly incomplete and only indicative but should nevertheless provide leads into the relevant literature for more information.

Metal coordination compounds, combining the numerous metal ions in their different oxidation states with the innumerable organic and inorganic ligands, offer a particularly fertile area of investigation and, indeed, their optical, photophysical, and photochemical properties have been actively studied.[2] Of special interest are those metal complexes that are highly polarizable due to specific ligands and/or present highly polarized and expanded excited states. *Mixed valence complexes* (possessing intervalence absorption bands) and *heterometallic binuclear complexes*, which undergo intramolecular electron transfer, display a variety of structural, physical, and chemical features of potential interest for optical studies, such are the bridged complexes $[(NH_3)_5RuN\!\!\bigcirc\!\!NRu(NH_3)_5]^{5+}$ (the Creutz–Taube ion) and $[(NH_3)_5Co^{III}N\!\!\bigcirc\!\!\bigcirc\!\!NFe^{II}(CN)_5]$, for instance.[3–5]

Charge transfer (CT) excited states in metal complexes are characterized by displacement of charge on photoexcitation either from the metal (M) to the ligand (L) ion (MLCT: metal to ligand charge transfer) or from the ligand to the metal ion (LMCT: ligand to metal charge transfer). Such states occur for instance in the complexes formed by polypyridines, phenanthroline, and their derivatives with various metal ions, like $Ru(bipy)_3^{2+}$, $Re(bipy)(CO_3)Cl$, or $Cu(phen)_2^+$ (bipy: α,α'-bipyridine; phen: phenanthroline), which have been very extensively studied.[5–8] In particular $Ru(bipy)_3^{2+}$ has an MLCT excited state $Ru(bipy)_3^{2+*}$ in which charge has been transferred from the central metal ion to the ligands, and which may be formulated as the expanded state $[Ru^{III}(bipy)_3^-]^{2+*}$. This transfer may be driven into a given direction by addition of electron accepting substituents A on a ligand unit, like in $[Ru^{III}(bipy)_2(bipyA_2)^-]^{2+*}$. The same holds for the MLCT state of $Re(bipy)(CO)_3Cl$ or of analogous Cu(I) complexes. Long-range $M \rightarrow L$ transfers may take place when acceptor groups are located far from the photosensitizer (PS) metal center.

Conversely, LMCT states occur in complexes like

$$Me_2N\!\!-\!\!\bigcirc\!\!NFe(CN)_5^{2-}$$

where, on excitation, charge is transferred away from the electron-donating dimethyl amino group toward the Fe^{III} center. Again, long-range transfer may be considered, for example in complexes like

Me_2N—⟨phenyl⟩—CH=CH—⟨pyridyl⟩—$NFe(CN)_5^{2-}$ O_2N—⟨phenyl, NO$_2$⟩—CH=CH—⟨pyridyl⟩—$NIrCl_5^{2-}$

LMCT →   ← MLCT

or with longer polyolefinic chains (such as carotenoids) separating the donor D or acceptor A groups from the PS metal center.

The strong polarization of such and other complexes in the excited state also makes them strongly *solvatochromic*, so that they may be used as color indicators for solvent parameters (e.g., $Cu(tmeda)(acad)NO_3$ or $Fe(phen)_2(CN)_2$; tmeda: tetramethylethylenediamine, acac: acetylacetonate).[9] This points to the possibility of modifying their optical properties in the solid state by ion pairing, which will be dependent on the nature of the counterions. Even stronger polarization is expected for *push-pull complexes* D———PS———A (of, for example, Ru(II), Re(I), Cu(I), etc.) incorporating both electron accepting and electron donating ligands, such as $[Ru(bipyD_2)_2(bipyA_2)]^{2+}$.

Another means of inducing optical changes rests on redox processes, since oxidation or reduction at the metal center in CT complexes [for instance $Fe^{II}(phen)_2(CN)_2$ and $Fe^{III}(phen)_2(CN)_2NO_3$][9], etc. should affect polarization and polarizability, shift absorption bands and could even reverse the direction of electron transfer on excitation. This would allow one to perform *electrochemical switching* of optical properties of metal complexes via redox changes.

One may point out that ion-dependent polarization occurs on complex formation with indicator ligands,[10] in which a dye residue is linked to a crown ether or cryptand[1,11] type binding site, thus giving rise to *ion-selective optical changes*. Although the complexing dyes developed as metal ion indicators are probably not ideal for optical studies, modifications involving mainly the introduction of other light-sensitive groups may lead to ligands and complexes displaying *cation control* of nonlinear optical properties.

An area of very active current research, which also is drawing heavily from the photochemistry of metal complexes, is that of *artificial photosynthesis*.[6-8,12] Its goal is to design chemical systems achieving the conversion and storage of light energy by photochemical processes such as water photolysis and carbon dioxide photoreduction. As in natural photosynthesis, the first step is to set up a reaction center achieving photoinduced charge separation, which then is to be used for driving catalytic reactions with sufficient efficiency to compete successfully with recombination. Suitable systems combine donor, acceptor, and photosensitizer components, and the process may be represented schematically as

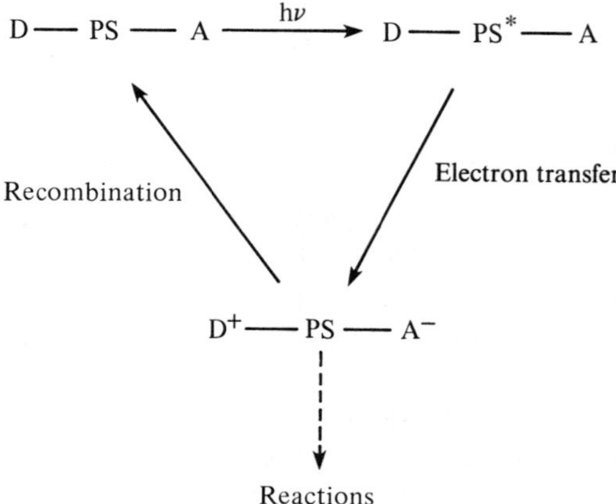

Thus, photosensitizers and photoprocesses developed for studies in artificial photosynthesis, may also be of interest for their optical properties. The main difference is probably that, whereas artificial photosynthesis requires the generation of long-lived charge-separated states, this is not the case for nonlinear optical properties.

Of course, other types of coordination compounds may possess photophysical properties of interest for nonlinear optics, for instance, one may cite lanthanides complexes,[13] complexes containing metal–metal bonds[14] and stacked extended arrays of metal complexes,[15] photoactive cryptates, complexes of metalloids, which have been comparatively very little investigated, etc.

Purely molecular complexes also offer a vast field of investigation. Thus, electron donor–acceptor (EDA) complexes (of $\sigma-\sigma$, $\sigma-\pi$, or $\pi-\pi$ type)[16a] possess polarized ground states and undergo (partial) intermolecular charge separation on excitation. Such is the case for donors like aromatic amines and acceptors like aromatic nitro compounds or quinones that form $\pi-\pi$ complexes where the flat donor and acceptor molecules are arranged in stacks. Other types include complexes between amines or aromatic molecules with halogens or halogenated molecules,[16b] complexes with Lewis acids derived from metalloids, complexes involving organic dye molecules, ion pairs possessing CT transitions such as those formed by pyridinium or bipyridinium salts, and complexes between heterocyclic donors and acceptors. Of great potential are also the actively studied EDA complexes, which possess properties of molecular semiconductors and conductors[17] as well as EDA derivatives of polymers. A particularly attractive feature of EDA complexes is the possibility to more or less finely tune their polarization, polarizability, extent of charge transfer, absorption bands, etc. by many variations in basic structural types as well as in substituents.

Numerous cases of biological EDA interactions involving chiral structures, in particular in proteins, are known, and the products resulting from the intercalation of organic dyes (like acridines) into helical nucleic acid strands represent a polymolecular case with multiple interactions.

This vast area of organic EDA complexes may also involve supermolecules formed by binding of D or A substrates respectively to A or D receptor molecules. Centrosymmetry can be lifted by using chiral components possessing asymmetric centers or atropisomerism. Highly polarizable stable free radicals (neutral or ionic) represent another intriguing type of molecular component. The possible structural types are too numerous for being considered here. Those mentioned are only intended to hint at the rich variety of the organic molecules available or still to be imagined and made, as well as of the supermolecules into which they may be combined.

Finally, on the *polymolecular level*, the substances of the various types mentioned may be incorporated into phases presenting a given supramolecular organization such as mesomorphic phases of various types, monolayers, molecular films, membranes, vesicles, etc.

The present very qualitative suggestions should be considered merely as incentives, and progress along these lines will evidently rest on actual realizations through design, synthesis, and optical studies.

It is hoped that these brief considerations on the pontential interest of intermolecular compounds, metal complexes, and molecular complexes as materials for investigations in nonlinear optics may suggest developments involving various types of supramolecular structures. Thus would be brought together intermolecular bonding and supramolecular architectures with optical properties of materials in a kind of *supramolecular photonics*.

REFERENCES

1. J. M. Lehn, *Science* **227**, 849 (1985).
2. V. Balzani and V. Carassiti, "Photochemistry of Coordination Compounds," Academic Press, New York, 1970.
3. H. Taube, *Pure Appl. Chem.* **41**, 25 (1975).
4. A. Haim, *Prog. Inorg. Chem.* **30**, 273 (1983).
5. T. J. Meyer, *Acc. Chem. Res.* **11**, 94 (1978).
6. V. Balzani, F. Bolletta, M. T. Gandolfi, and M. Maestri, *Top. Curr. Chem.* **75**, 1 (1978).
7. N. Sutin, *J. Photochem.* **10**, 19 (1979); N. Sutin and C. Creutz, *Pure Appl. Chem.* **52**, 2717 (1980).
8. K. Kalyanasundaram, *Coord. Chem. Rev.* **46**, 159 (1982).

9. R. W. Soukup and R. Schmid, *J. Chem. Educ.* **62,** 459 (1985); R. E. Shepperd, M. F. Hog, N. Hoblack, and C. R. Johnson, *Inorg. Chem.* **23,** 3249 (1984) and references therein.
10. M. Takagi and K. Ueno, *Top. Curr. Chem.* **121,** 39 (1984).
11. J. M. Lehn, *Acc. Chem. Res.* **11,** 49 (1978).
12. (a) J. Connolly (ed.), "Photochemical Conversion and Storage of Solar Energy," Academic Press, New York, 1981; (b) M. Grätzel (ed.), "Energy Resources through Photochemistry and Catalysis," Academic Press New York, 1983; (c) J. M. Lehn, Chapter 6 in reference (a) and *Proc. Int. Congr. Catal., 8th, Berlin,* Vol. 1, p. 63. Verlag Chemie, Weinheim, 1984.
13. J. C. G. Bunzli and D. Wessner, *Coord. Chem. Rev.* **60,** 191 (1984).
14. T. J. Meyer and J. V. Caspar, *Chem. Rev.* **85,** 187 (1985).
15. J. S. Miller and A. J. Epstein, *Prog. Inorg. Chem.* **20,** 1 (1976); C. Piechocki and J. Simon, *Nouv. J. Chim.* **9,** 159 (1985).
16. (a) H. A. Bent, *Chem. Rev.* **68,** 587 (1968); R. Foster, "Organic Charge-Transfer Complexes," Academic Press, New York, 1969; R. Foster (ed.), "Molecular Complexes," Vols. 1 and 2, Elek Science, London; Crane Russak, New York, 1973 and 1974; (b) For a relevant study see: B. F. Levine and C. G. Bethea, *J. Chem. Phys.* **65,** 2439 (1976).
17. H. Meier, "Organic Semiconductors," Verlag Chemie, Weinheim, 1974; J. Simon and J.–J. André, "Molecular Semiconductors," Springer–Verlag, Berlin and New York, 1985.

Appendix I

Organic SHG Powder Test Data

J. F. NICOUD and R. J. TWIEG

This appendix is a compilation of optical second-harmonic generation powder efficiency measurements on a wide range of organic compounds. The primary literature references [1–73] are given in chronological order (1964–1985), and a number of other sources, including reviews [74, 75], books [76–78], dissertations, [79, 80], other miscellaneous sources [81–90] and references added in proof [91–97] are also provided. Due to space considerations, the appendix is not comprehensive, but it does contain representative compounds from all of the primary references (which should be further consulted for additional examples). The focus of this appendix is on organic molecular compounds only, and although a number of organic salts and polymeric materials are included, no particular effort has been made to identify these materials and include them. The Landolt–Börnstein volumes [81, 82, 97] are of particular value for data on conventional inorganic ionic nonlinear materials and are also useful for the organic salts and even a few fully organic compounds.

The first systematic investigation of the powder measurement technique is due to Kurtz and Perry [6]. Their effort involved the design of an experimental apparatus and the theoretical treatment of experimental parameters (beam incidence angle θ, sample thickness L, average particle size $\hat{r}$ and beam diameter D) that determine the intensity of the second harmonic

$I^{2\omega}$. In addition, a five-category material classification scheme was devised, and a number of novel and important nonlinear organic molecules were discovered as well. Since the initial effort of Kurtz and Perry, some further enhancements have been made in this important experimental technique [93, 94].

This appendix is organized as a function of molecular structure and functionality into eight general tables, which are further subdivided when convenient. This organization is arbitrary but offers a means to provide some degree of structure–activity correlation (other equally valid organizational routines might have utilized efficiency or chromophore as a primary sorting mechanism). In cases where the molecule is multifunctional, it was placed according to what is felt to be the largest functional contribution to the molecular hyperpolarizability. For example, N-(4-nitrophenyl)-(S)-prolinol (NPP) is included in the nitroaromatic table (IV,A,2,a) rather than in the amino acid derivative table (II), since the large hyperpolarizability of this molecule clearly resides in the nitroaniline functionality. As examples of some less clear-cut cases, diphenylurea is included in the urea table (III,A), since it is likely that the largest contribution to the molecular hyperpolarizability resides in the urea functionality rather than the aromatic part of the molecule. However, dinitrodiphenylurea (DNPU) is included in the nitroaromatic table (IV,A,2,a), since the largest component of the molecular hyperpolarizability resides in the nitroaromatic part of the molecule and the contribution from the urea portion is probably relatively insignificant. These choices involving structural segregation are in many cases arbitrary and should not be construed to be experimentally determined or even particularly well-founded. Nonetheless, some sort of segregation is mandated.

In terms of data, this appendix first provides a chemical and/or trivial name for the compound (of which neither is necessarily unique). In cases where the point group (or space group) for a compound is known, it has been provided. The literature references for the crystallographic data are not provided except for a few specific primary references [24, 34, 46] in which the crystallographic paper specifically discusses the nonlinear optical properties of the compound. Detailed crystallographic data can be found in the original publication or in a compilation such as the Cambridge file. The typical powder efficiency (or range thereof) is provided relative to some standard (described in the cross-reference section at the end of the appendix), and any single-crystal data (such as d values in units of pmV^{-1}) for the material are provided if available. The powder efficiencies reported must be interpreted with caution. The powder efficiency obtained on a specific specimen is a function of a host of parameters, inherent in that particular specimen and experimental design, that are very often neither controlled nor reported. The sample history and purity are important, since quite often a single substance will have a variety of crystal modifications with different optical properties and a particular specimen may be a pure modification or a mixture of modifications. The modification is often determined by the recrystallization solvent, which itself may be cocrystallized or occluded. For optically active substances, the situation is even more complicated, since the extent of optical purity (enantiomeric excess) will influence the efficiency. It is assumed that the optically active compounds (indicated by a D, L, R, S or an asterisk in the tables) are optically pure, but this criterion is rarely specified. Other experimental parameters include dispersion, particle size distribution, phase-matching properties, and temperature [91, 94]. The effect of dispersion has not been well quantified on powder samples, but in some cases, especially when the material's absorption edge is

TABLE I

Saturated Compounds (No π Bonds)

Compound	Crystal data	Powder efficiency or d [pmV^{-1}]	Reference
A. Simple alcohols, amines, etc.			
Menthol*	3	1.5Q	[6, 67]
Cholesterol* (Note $\Delta^{5,6}\pi$-bond)	1	1Q	[6]
Hexamethylenetetramine	$I\bar{4}3m$	10K	[2]
B. Carbohydrates			
D-Sucrose	2	0.2K	[3, 59, 68]
D-Glucose	222	0.1K 20Q	[3] [6]
β-D-Lactose	2	20Q 5 × Sucrose	[6] [68]

near the second harmonic, it may be important. In relation to this, the use of a tunable source offers the opportunity to test the phase matching properties of optical materials in the powder form [52]. In a number of cases [11, 22, 72, 93, 94] the relationship between powder efficiency and particle size has been determined. This relationship is strongly material-dependent, and a change of particle size by a factor of 10 may change the resultant powder efficiency little or by as much as an order of magnitude. In certain cases when the crystal cleaves in a fashion to give a large face that is phase-matchable, anomalous results in powder efficiency of up to an order of magnitude have been found [15].

Given these complications and the large number of independent sources for the powder efficiency data, each value should not be taken to be accurate to anything less than a factor of two (it has been stated that the powder technique is no more accurate than a factor of five) [15]. In spite of the complications, with these few caveats in mind the powder technique remains an extremely valuable tool for initial screening of organic materials for second-harmonic generation.

TABLE II

Amino Acids and Derivatives

Compound		Crystal data	Powder efficiency or d [pmV^{-1}]	Reference
D,L,DL Forms of 24 amino acids and amino acid hydrochlorides			0.0–1.7K	[3, 5]
Hippuric acid	$C_6H_5CONHCH_2CO_2H$	222	5ADP	[4]
L-Histidine		222	10Q	[6]
L-Glutamic acid	$HO_2CCH_2CH_2CH(NH_2)CO_2H$	222	8Q	[6]
L-Tryptophan			0.3mNA	[32]
L-Asparagine	$H_2NCOCH_2CH(NH_2)CO_2H$	$P2_12_12_1$	0.3mNA	[32]
L-Methionine sulfone	$CH_3SO_2CH_2CH_2CH(NH_2)COOH$		0.3L	[17]
N-Acetyl-L-cysteine	$HSCH_2CH(NHCOCH_3)CO_2H$	1	0.2U	[88]
71 Amino acids, assorted peptides, and proteins			0.0–7.0K	[42]

TABLE III

Ureas and Thioureas

Compound		Crystal data	Powder efficiency or d [pmV^{-1}]	Reference
A. Urea and derivatives				
Urea	H_2NCONH_2	$\bar{4}2m$	400Q	[6, 35, 38, 40, 41, 45, 67, 96]
Diphenylurea		$mm2$	0.2mNA	[18, 23]
sym-Dimethylurea	$CH_3NHCONHCH_3$	$mm2$	2K	[44, 62]
			0.3–3U	[52]
Bispentamethyleneurea			1U	[52]
Formylurea	$HCONHCONH_2$		0.15U	[54]
1,1'-Methylenediurea	$H_2NCONH—CH_2—NHCONH_2$		0.35U	[54]
4-Chlorophenylurea	$ClC_6H_4NHCONH_2$		0.35U	[54]
N-Nitrourea	$O_2N—NHCONH_2$		4K	[62]
B. Thiourea and derivatives				
Thiourea	H_2NCSNH_2	$mm2$	Weak	[80, 83]
N-(α-Methylbenzylamino)- N'-phenylthiourea*			0.3L 0.5L	[9] [17]
N-(α-Methylbenzylamino)- N'-dimethylthiourea*	$C_6H_5CH(CH_3)NHCSN(CH_3)_2$		0.3L	[17]
1-Isonicotinyl-3-thiosemicarbazide			0.35U	[54]
Dimethylthioparabanic acid			4U	[88]

TABLE IV

Aromatic Compounds

Compound	Crystal data	Powder efficiency or d [pmV^{-1}]	Reference
A. Derivatives of benzene			
1. With saturated substituents			
Resorcinol (1,3-dihydroxybenzene)	$mm2$ $Pbn2_1$	20Q 0.03mNA $d_{311} = 2.3$	[6, 10, 16, 21, 36]
1,3-Diodobenzene	$mm2$	$d_{32} = 3.3$	[10]
3-Aminophenol	$mm2$	0.2mNA	[16, 17, 28, 31]
p-Anisidine	$P2_1/c$ $Pmc2_1$	0.1mNA	[18, 22]
3-Methyl-4-isopropylphenol	$P4_1$	$d = d(K)$	[33]
m-Toluenediamine (MTDA)	$Pna2_1$	$d = 2.9d(K)$	[29, 46, 80]
Tetraphenylmethane (silane, tin, lead)	$\bar{4}2m$	0.1U	[86] [80]
2. With conjugating (π-bond-containing) substituents			
a. Nitro			
1,3-Dinitrobenzene	$mm2$ $Pbn2$	500Q 1L 0.4mNA $d_{31} = 4.2$	[6, 10, 11, 13, 16, 27, 37, 95]

Compound	Space group		References
3-Nitroaniline (mNA)	$Pbc2_1$	1L $d_{333} = 17.6$ $d_{311} = 15.1$ $d_{322} = 1.7$	[9, 11–13, 16–19, 21, 24, 26–28, 31, 32, 36, 37, 48]
4-Nitrophthalimide		0.1mNA	[11]
N-α-Methylbenzylamino- N'-(4-nitrophenyl)thiourea*		1L, 2L 2mNA	[13, 17] [21]
4-Dimethylaminonitrobenzene	$P2_1$	0.5mNA	[18, 22]
2,4-Dinitrophenol	222	0.6mNA	[18, 76]
3-Nitrobenzaldehyde		0.7mNA	[18, 21, 22, 76]
4-Nitrobenzaldehyde		0.3mNA 1U 0.2mNA	[18, 21, 76, 88]
4-Nitrophenylhydrazine (NPH)		1mNA 1.6mNA	[19] [28, 91, 92]
2-Chloro-4-nitroaniline	$mm2$	40U	[16, 27, 87]
2-Bromo-4-nitroaniline	$mm2$	3U	[16, 27, 87]
1-Formyl-2-(4-nitrophenyl hydrazone) (FNPH)		1mNA	[28]
2-Methyl-3-nitroaniline		0.5mNA	[28]

(*continues*)

TABLE IV (*Continued*)

Compound	Crystal data	Powder efficiency or d [pmV^{-1}]	Reference
3-Nitrobenzonitrile		0.4mNA	[32]
		1U	[88]
4-Nitrobenzonitrile	$P2_1$	0.8mNA	[32]
		2U	[88]
		2U	[84]
4-Nitrophenol		0.7mNA	[28]
Acetophenone-4-nitrophenyl-hydrazone		0.8mNA	[28]
Benzaldehyde-4-nitrophenyl-hydrazone		0.6mNA	[28]
Camphor-4-nitrophenyl hydrazone*		3U	[84]
Carvone-4-nitrophenyl hydrazone*		10U	[84]
2,4-Dinitro-2′-methoxy-diphenylamine		0.4mNA	[28]
3-Nitrobenzaldehyde semicarbazone		0.8mNA	[32]
4-Nitrobenzaldehyde hydrazine		1mNA	[32]

Compound		Space group	Value	Ref.
2-Methyl-4-nitroaniline (MNA)	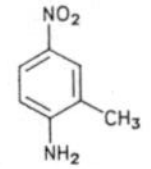	Cc	$d_{12} = 38$ $d_{11} = 250$	[39, 50, 51, 76, 80]
4-Nitro-4′-methylbenzylidene aniline (NMBA)	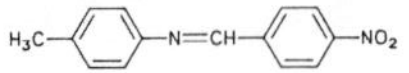	Pb	$d = 16.7$	[34]
4 methoxy-4′-nitrostilbene	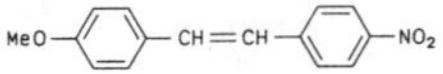		10–2500Q	[94]
2,4-Dinitrophenyl-(L)-serine	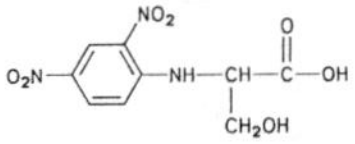		1.0mNA	[32]
2,4-Dinitrophenyl-(L)-alanine	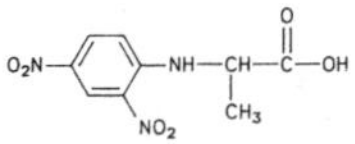		1mNA	[32]
2,4-Dinitrophenyl-(L)-alanine methylester (MAP)	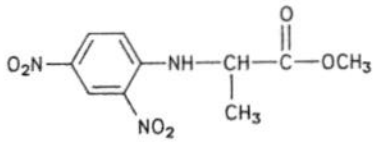	$P2_1$	$d_{22} = 18.4$ 10U	[30, 57]
4-Nitrophenyl-(L)-alanine methylester	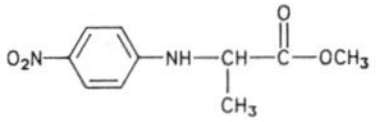		1U	[54, 57]
Various amino acid ester derivatives of 2,4-dinitrobenzene	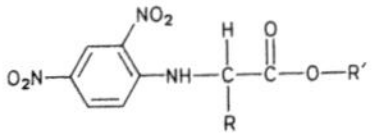		5–3000Q	[58, 79]
(2,4-Dinitro-5-fluoro)phenyl-(L)-alanine methylester	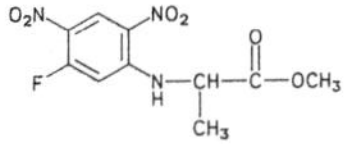		21U	[85]
(2,4-Dinitro-5-chloro)phenyl-(L)-alanine methylester	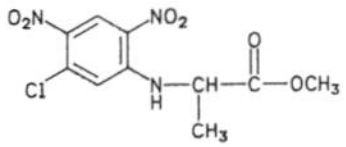		2U	[85]
3-Nitro-N-methylaniline	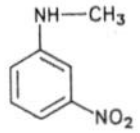		1mNA	[17]

(*continues*)

TABLE IV (*Continued*)

Compound	Crystal data	Powder efficiency or d $[\mathrm{pmV}^{-1}]$	Reference
2,4-Dinitrophenylphthalimide	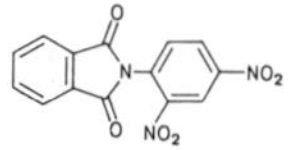	1mNA	[32]
3-Nitrophenylphthalimide	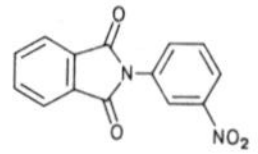	0.5mNA	[32]
3,3′-Dinitrobiphenyl	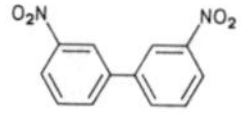	0.2mNA 0.6U	[32] [88]
2,4-Dinitrophenylsemicarbazide (DNP-SC)	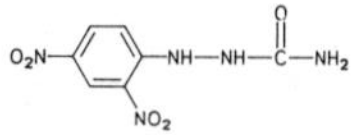	8.8U	[53, 54]
Di-(*p*-nitrophenylurea) (DNPU)	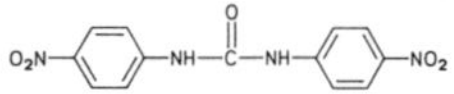	8.8U	[53, 54, 66]
4-Amino-4′-nitrobiphenyl	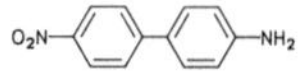	0.6mNA	[32]
m-Nitroisophthalic acid	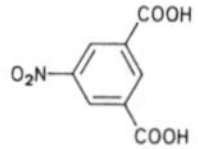	0.1mNA	[32]
2,4-Dinitro-6-chloroaniline	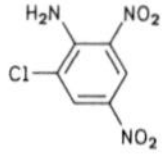	0.6mNA	[32]
1,8-Dinitronaphthalene	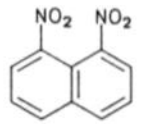 $P2_12_12_1$	<0.1mNA	[18]

Compound		Value	Ref.
2,4-Dinitro-6-chlorophenol		0.5mNA	[76]
2,5-Dinitrofluorene		6K	[62]
2,4,5,7-Tetranitrofluorenone		2K	[62]
N-(2,4-Dinitrophenyl)-p-toluidine		2K	[62]
N-(2,4-Dinitrophenyl)-m-toluidine		3K	[62]
2,4-Dinitrophenylhydrazine		3K	[62]
N-(n-Butyl)-2,4-dinitroaniline		6K	[62]
N-(2,4-Dinitrophenyl)-N'-tosyl-p-phenylenediamine		10K	[62]
2-Acetamido-4,5-dinitroaniline		20K	[62]
2-Methyl-4-nitro-N-methylaniline (MNMA)	$Pna2_1$	10K 80U	[62] [85]
2-Methoxy-4-nitro-N-methylaniline		15K	[62]

(continues)

TABLE IV (*Continued*)

Compound	Crystal data	Powder efficiency or d [pmV^{-1}]	Reference
3-Acetamido-4-aminonitro-benzene		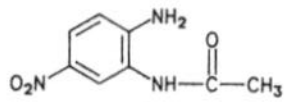 20U	[85]
3-Propionamido-4-amino-nitrobenzene		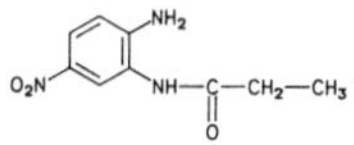 10U	[85]
3-Propionamido-4-methylamino-nitrobenzene		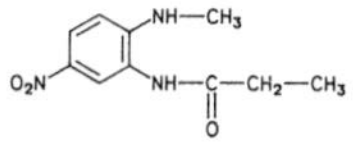 5U	[85]
3-Acetamido-4-pyrrolidino-nitrobenzene (PAN)	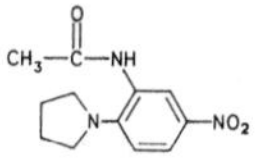	80U	[85]
3-Acetamido-4-dimethylamino-nitrobenzene (DAN)	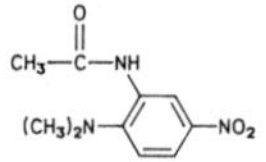 $P2_1$	115U	[85]
3-Trifluoroacetamido-4-dimethyl-aminonitrobenzene		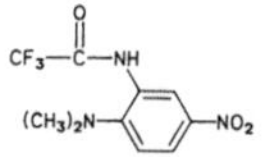 70U	[85]
2,4-Bis(phenylthio)nitrobenzene	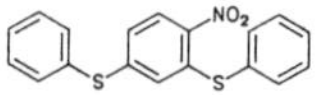	4U	[88]
4-*N*-(Cycloheptylamino)nitrobenzene	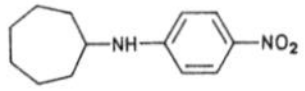	7U	[87]
4-nitrocatechol	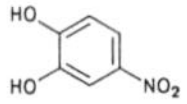	6U	[88]

232

Compound	Structure		Ref.
2,3,5,6-Tetrafluoro-4-*N*-(α-methylbenzylamino)-nitrobenzene*	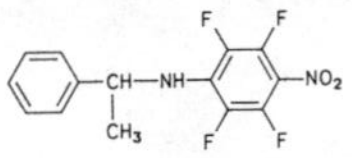	15U	[83]
4-Nitrobenzylchloride	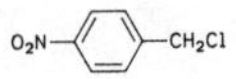	1U	[88]
N-(4-Nitrophenyl)-*N'*-(α-methylbenzylamino)urea*	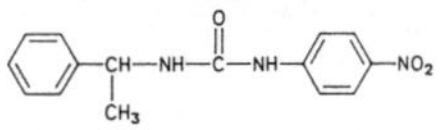	12U	[83]
2-Chloro-4-nitro-*N*-methylaniline	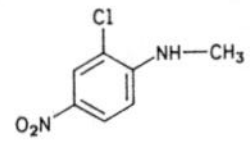	20U	[87]
N-(4-Nitrophenyl)arabinisoyl pyranosylamine*	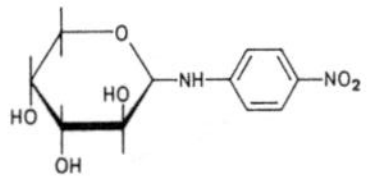	10U	[89]
3-Methyl-4-*N*-(α-methylbenzylamino)-nitrobenzene*	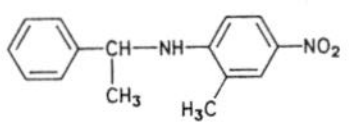	6U	[85]
N-(2,4-Dinitrophenyl)-1-(1-naphthyl)-ethylamine*	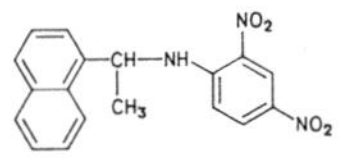	16U	[85]
N-(2,4-Dinitro-5-fluorophenyl)-1-(1-naphthyl)ethylamine*	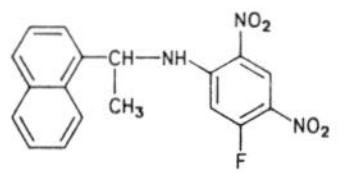	10U	[85]
N-(2,4-Dinitro-5-chlorophenyl)-1-(1-naphthyl)ethylamine*	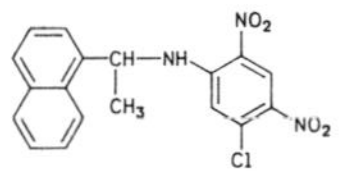	17U	[85]

(continues)

233

TABLE IV (*Continued*)

Compound		Crystal data	Powder efficiency or d [pmV^{-1}]	Reference
N-(2-Methylbutyl)-4-nitroaniline*			15U	[84]
N-(4-Nitrophenyl)-(s)-alaninol			15U	[84]
N-(2-Methyl-4-nitrophenyl)-(s)-alaninol			7U	[84]
N-(2-Cyano-4-nitrophenyl)-(s)-alaninol			7U	[84]
N-(4-Nitrophenyl)-2-amino-1-butanol*			8U	[84]
N-(4-Nitrophenyl)-1-amino-2-propanol*			10U	[84]
N-(4-Nitrophenyl)-pseudoephedrine*			15U	[84]
N-(4-Nitrophenyl)-nor-pseudo-ephedrine*			6U	[84]
N-(4-Nitrophenyl)-(s)-prolinol (NPP)		$P2_1$	150U	[72]

Compound		Value	Ref.
N-(2-Cyano-4-nitrophenyl)-(s)-prolinol	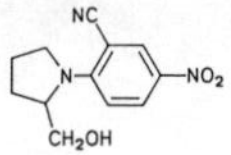	15U	[84]
N-(3-Hydroxy-4-nitrophenyl)-(s)-prolinol (HNPP)	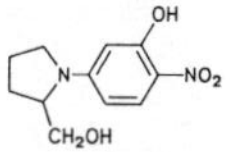	140U	[84]
N-(2,4-Dinitro-5-fluorophenyl)-(s)-prolinol	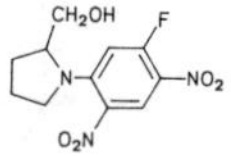	3U	[85]
N-(2,4-Dinitro-5-chlorophenyl)-(s)-prolinol	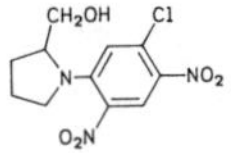	9U	[85]
N-(3-Methyl-4-nitrophenyl)-(s)-prolinol	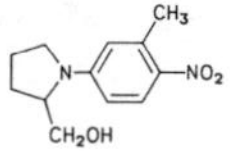	3U	[84]
N-(4-Nitrophenyl)-trans-1,2-diaminocyclohexane*	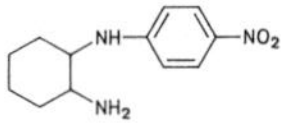	7U	[84]
1-(4-Nitrophenyl)-2-anilinomethyl-pyrrolidine*	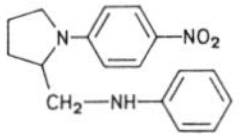	6U	[84]
N-2-(2-Hydroxy-1-aminoethyl)-5-nitrobenzoic acid	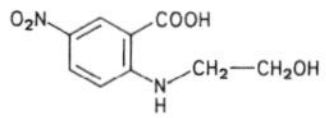	10U	[84]
N-(4-Nitrophenyl)-3-amino-1-propanol (APNP)	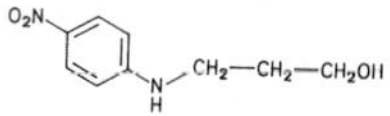	80U	[84]

(continues)

TABLE IV (*Continued*)

Compound	Crystal data	Powder efficiency or d $[\text{pmV}^{-1}]$	Reference
N-(2-Methyl-4-nitrophenyl)-3-amino-1-propanol		6U	[84]
N-(2-Trifluoromethyl-4-nitrophenyl)-3-amino-1-propanol		30U	[84]
N-(3-Trifluoromethyl-4-nitrophenyl)-3-amino-1-propanol		80U	[84]
N-(3-Trifluoromethyl-4-nitrophenyl)-4-Hydroxypiperidine		12U	[84]
N-(2-Cyano-4-nitrophenyl)-4-hydroxypiperidine		6U	[84]
N-(4-Nitrophenyl)-*N*-methyl-2-aminopropionitrile (NPPN)		85U	[84]
N-(4-Nitrophenyl)-*N*-methyl-2-aminoacetonitrile (NPAN)	Fdd2	140U	[84]
N-(4-Nitrophenyl)-4-aminobutanoic acid (BANP)		115U	[84]
b. Carbonyl (acid, ester, amide, etc.)			
3-Aminophthalimide		3K	[8]

Compound	Structure	Space group		Reference
Benzil	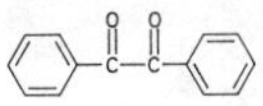	$P32$	$d_{11} = 0.8$	[10]
Benzophenone	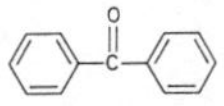	$P2_12_12_1$	$d_{36} = 0.4$	[10, 23]
Bis-4,4′-(2-hydroxymethyl-pyrrolidino)-benzophenone*	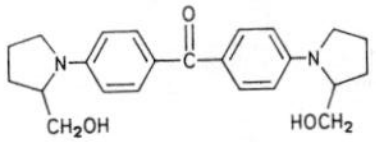		5U	[84]
Methyl 4-hydroxybenzoate	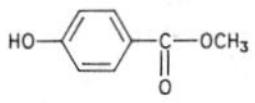		0.5mNA	[11]
Phthalic anhydride	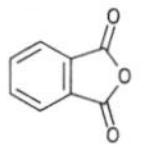	$Pna2_1$	0.3mNA	[11, 18, 21, 23]
Ethyl 4-aminobenzoate	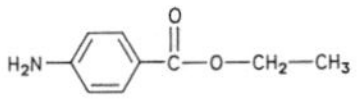	$P2_12_12_1$	0.8mNA	[18, 21, 22]
2,6-Di-*t*-butyl-4-hydroxy-benzaldehyde	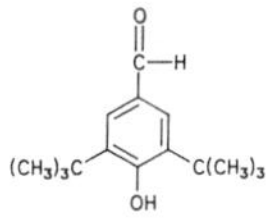		0.7mNA 8U	[32] [83]
N,N-Dimethyl-4-bromobenzamide	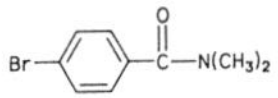	$Pna2_1$	0.4mNA	[32]
3-Methoxy-4-hydroxybenzaldehyde	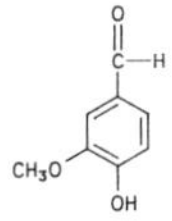		1mNA	[32]
4-Aminobenzaldehyde	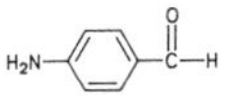		0.5mNA	[32]

(*continues*)

TABLE IV (*Continued*)

Compound		Crystal data	Powder efficiency or d $[pmV^{-1}]$	Reference
Diphenic anhydride	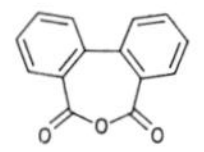		0.3mNA	[32]
3-Hydroxybenzoic acid	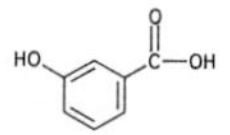		0.4mNA	[32]
1,4-Naphthoquinone	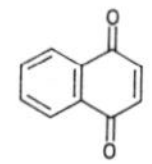		0.4mNA	[28]
2-Methyl-*p*-quinone	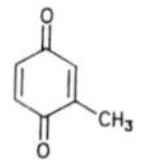		2K	[62]
4-Bromodibenzoylmethane	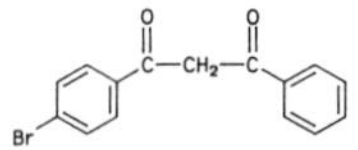		2mNA	[76]
4,4′-Dibromodibenzoylmethane	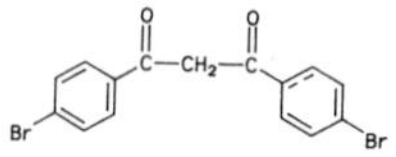		4mNA	[76]
4-Bromoacetophenone	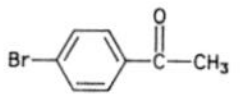		1U	[83]

Compound	Structure	Space group		Reference
4-Aminobenzamide		$P2_1$	2U	[84, 88]
Tribenzoylmethane			7K	[62]
c. Sulfur derivatives (sulfone, sulfonamide, sulfonate, etc.)				
Sulfanilic acid			5K	[62]
4-Trifluoromethylsulfonylaniline			0.5mNA	[76]
4,4′-Dihydroxydiphenylsulfone			0.3mNA	[32]
4,4′-Diaminodiphenylsulfone			0.2mNA	[32]
4-Methylphenyl-β-styrene sulfonate			2K	[62]
Phenyl 3-nitrobenzenesulfonate			1K	[62]
Ethylsulfone 3-nitroanilide			2K	[62]
d. Other				
1,2-Dicyanobenzene (phthalonitrile)			0.6mNA 1U	[32] [88]

(continues)

TABLE IV (*Continued*)

Compound		Crystal data	Powder efficiency or d [pmV^{-1}]	Reference
Triphenylbenzene	(C$_6$H$_5$)$_3$C$_6$H$_3$	*Pna*2$_1$	$d_{333} = 1.0$	[36]
3-Acetylaminobenzonitrile			0.2mNA	[32]
4-Cyanophenyl-(s)-prolinol			2U	[83]
B. Derivatives of pyridine 1. Pyridine *N*-oxide 4-Nitropyridine *N*-oxide			Weak	[18]
3-Methyl-4-nitropyridine *N*-oxide (POM)		*P*2$_1$2$_1$2$_1$	13U $d = 9.2$	[49, 57, 63]
3-Chloro-4-nitropyridine *N*-oxide			Medium	[49]
3-Bromo-4-nitropyridine *N*-oxide			Weak	[49]

Compound			
4-Acetylpyridine *N*-oxide	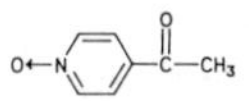	Weak	[49]
2. Nitropyridines			
2-Chloro-5-nitropyridine	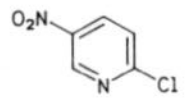	1.5U	[83, 84]
2-Chloro-3,5-dinitropyridine	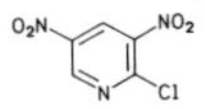	8U	[57, 85]
2-Phenoxy-3,5-dinitropyridine	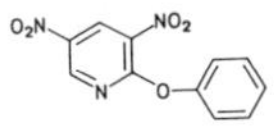	8U	[57, 85]
2-Hydroxy-3,5-dinitropyridine	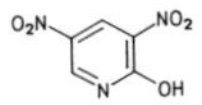	5U	[57, 85]
2-*N*-(α-methylbenzylamino)-3,5-dinitropyridine (MBADNP)*	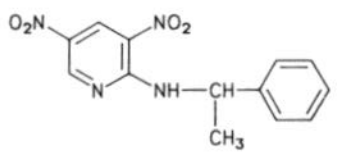 $P2_1$	10U	[55, 57, 85]
2-*N*-(α-methylbenzylamino)-5-nitropyridine (MBANP)*	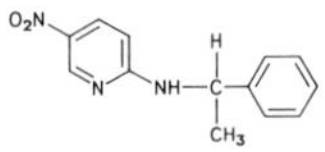 $P2_1$	25U	[55, 57, 85]
2-*N*-(α-Methylbenzylamino)-3-methyl-5-nitropyridine*	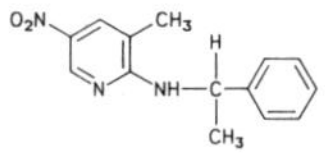	8U	[55, 57, 85]
2-*N*[α-(1-Ethylnaphthyl)amino]-3-methyl-5-nitropyridine*	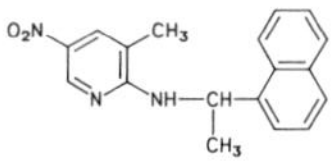	25U	[55, 57, 85]

(continues)

TABLE VII

Salts

Compound		Crystal data	Powder efficiency	Reference
(L)-Lysine hydrochloride	$H_2N(CH_2)_4CH(NH_2)CO_2H \cdot HCl$	2	<100Q	[6]
(L)-Arginine hydrochloride	$H_2NC(=NH)NH(CH_2)_3CH(NH_2)CO_2H \cdot HCl$	$4P2_1$	0.3mNA	[32]
Potassium (L)-aspartate	$HO_2CCH_2CH(NH_2)CO_2K$		0.3mNA	[32]
Ammonium malate, hydrate (AM)	$HO_2CCH_2CH(OH)CO_2NH_4 \cdot xH_2O$	Cs	1K	[38]
Potassium malate, hydrate (KM)	$HO_2CCH_2CH(OH)CO_2K \cdot xH_2O$		$d_{31} = 1.4d_{36}(K)$	[61]
Lanthanide formates	$Ln(HCO_2)_3$	$R3m$	1K	[43, 62]
Yttrium formate dihydrate	$Y(HCO_2)_3 \cdot 2H_2O$	$P2_12_12_1$	0.5K	[65]
Yttrium formate anhydrous	$Y(HCO_2)_3$	$R3m$	2K	[70]
Sodium p-nitrophenolate hydrate	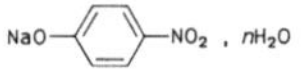 $NaO-\!\!\!\bigcirc\!\!\!-NO_2 , nH_2O$		1mNA	[32]
Lithium vanillinate, hydrate	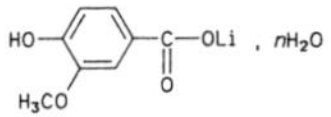 $HO-\!\!\!\bigcirc\!\!\!-C(=O)OLi , nH_2O$ (with H_3CO)		0.4mNA	[32]
Potassium 4-aminobenzene sulfonate	 $H_2N-\!\!\!\bigcirc\!\!\!-SO_3K$		1K	[62]
Sodium 3-nitrobenzene sulfonate	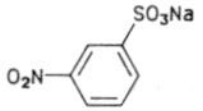 $O_2N-\!\!\!\bigcirc\!\!\!-SO_3Na$		1K	[62]
(L)-Arginine phosphate monohydrate (LAP)	$H_2NC(=NH)NH(CH_2)_3CH(NH_2)CO_2H \cdot H_3PO_4 \cdot H_2O$	$P2_1$	3.5K	[64]
Pyrrolidinium pyrrolidine dithiocarbamate	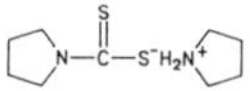 $N-C(=S)-S^-H_2N^+$ (pyrrolidine rings)	$P2_1$	1U	[87]
4-Dimethylamino-N-methyl-4-stilbazolium salts, $CH_3OSO_3^-$, ReO_4^-, BF_4^-	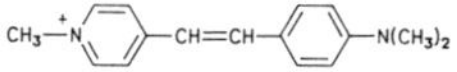 $CH_3-N^+\!\!\!\bigcirc\!\!\!-CH=CH-\!\!\!\bigcirc\!\!\!-N(CH_3)_2$	$Cmc2_1$	30mNA 18mNA 10mNA	[56] [73]

TABLE VIII

CT Complexes, Mixed Crystals, Inclusion Compounds

Compound		Crystal data	Powder efficiency	Reference
Acenaphthene picrate			0.4mNA	[28]
4-Nitroaniline/4-nitrophenol, 1:1			6K	[62]
4-Nitroaniline/dimethyl-β-cyclodextrin*			4U	[71]
Urea/resorcinol, 1:1		$P2_12_12_1$	1U	[84]

UNITS

d_{ij} SHG coefficient or second-order nonlinear dielectric susceptibility in pmV^{-1}

β Molecular hyperpolarizability in $m^4\,V^{-1}$.

CONVERSIONS

d_{ij} $1\,pmV^{-1} = \dfrac{3 \times 10^{-8}}{3\pi}\ \mathrm{esu}(\mathrm{erg\ cm^{-3}})^{-1/2} = 2.387 \times 10^{-9}\ \mathrm{esu}$

β $\beta\ (SI) = \dfrac{4\pi}{3 \times 10^{10}}\,\beta\ (CGS) = 4.1888 \times 10^{-10}\,\beta\ (CGS)$

CROSS REFERENCES

Symbol	Substance	Formula	$d\,[pmV^{-1}]$
K	KDP	KH_2PO_4	$d_{14} = 0.67, d_{36} = 0.63$
L	Lithium niobate	$LiNbO_3$	$d_{31} = -5.95$
A	ADP	$NH_4H_2PO_4$	$d_{36} = 0.76$
Q	Quartz	SiO_2	$d_{11} = 0.50$
U	Urea	NH_2CONH_2	$d_{14} = 2.3$
mNA	m-Nitroaniline	$1NH_2 \cdot 3NO_2 \cdot C_6H_4$	$d_{333} = 17.6, d_{311} = 15.1$

Approximate conversions among powdered samples: U = 0.4L = 3K = 20–70Q; 1mNA = 1L; 1A = 1.1K. These are typical powder sample conversions from a variety of literature sources; note that they do not necessarily scale as the relative d_{ij} values.

REFERENCES

1. P. M. Rentzepis and Y. H. Pao, *Appl. Phys. Lett.* **5** (8), 156 (1964).
2. G. H. Heilmeier, N. Ockman, R. Braunstein, and D. A. Kramer, *Appl. Phys. Lett.* **5** (11), 234 (1964).
3. K. Rieckhoff and W. F. Peticolas, *Science* **147**, 610 (1965).
4. R. Y. Orlov, *Sov. Phys.–Crystallogr. (Engl. Transl.)* **11** (3), 410 (1966).
5. V. S. Suvorov and A. S. Sonin, *Sov. Phys.–Crystallogr. (Engl. Transl.)* **11** (5), 711 (1967).
6. S. K. Kurtz and T. T. Perry, *J. Appl. Phys.* **39** (8), 3798 (1968).
7. M. Bass, D. Bua, R. Mozzi, and R. R. Monchamp, *Appl. Phys. Lett.* **15** (12), 393 (1969).
8. L. D. Derkecheva, A. I. Krymova, and N. P. Sopina, *JETP Lett. (Engl. Transl.)* **11** (10), 319 (1970).
9. B. L. Davydov, L. D. Derkacheva, V. V. Dunina, M. E. Zhabotinskii, V. F. Zolin, L. G. Koreneva, and M. A. Samokina, *JETP Lett. (Engl. Transl.)* **12** (1), 16 (1970).
10. J. R. Gott, *J. Phys. B* **4**, 116 (1971).
11. J. Jerphagnon, *IEEE J. Quantum Electron.* **QE-7**, 42 (1971).
12. P. D. Southgate and D. S. Hall, *Appl. Phys. Lett.* **18** (10), 456 (1971).
13. B. L. Davydov, L. D. Derkacheva, V. V. Dunina, M. E. Zhabotinskii, V. F. Zolin, L. G. Koreneva, and M. A. Samokhina, *Opt. Spectrosc. (Engl. Transl.)* **30**, 274 (1971).
14. P. D. Southgate and D. S. Hall, *J. Appl. Phys.* **42** (11), 4480 (1971).
15. J. G. Bergman, G. R. Crane, B. F. Levine, and C. G. Bethea, *Appl. Phys. Lett.* **20** (1), 21 (1972).
16. P. D. Southgate and D. S. Hall, *J. Appl. Phys.* **43** (6), 2765 (1972).
17. B. L. Davydov, V. V. Dunina, V. F. Zolin, L. G. Koreneva, M. A. Samokhina, and E. P. Shliteris, *Opt. Spectrosc. (Engl. Transl.)* **32**, 118 (1972).
18. B. L. Davydov, V. F. Zolin, L. G. Koreneva, and M. A. Samokhina, *J. Appl. Spectrosc.* **17** (3), 1132 (1972).
19. G. P. Bolognesi, S. Mezzetti, and F. Pandarese, *Opt. Commun.* **8** (3), 267 (1973).
20. V. D. Shigorin and G. P. Shipulo, *Opt. Spectrosc. (Engl. Transl.)* **34** (1), 83 (1973).
21. B. L. Davydov, V. V. Dunina, V. F. Zolin, and L. G. Koreneva, *Opt. Spectrosc. (Engl. Transl.)* **34** (2), 150 (1973).
22. B. L. Davydov, V. F. Zolin, L. G. Koreneva, and M. A. Samokhina, *J. Appl. Spectrosc.* **18** (1), 120 (1973).
23. V. D. Shigorin and G. P. Shipulo, *Sov. Phys.–Crystallogr. (Engl. Transl.)* **18** (3), 349 (1973).
24. A. C. Skapski and J. L. Stevenson, *J. Chem. Soc., Perkin Trans. 2*, p. 1197 (1973).
25. B. L. Davydov, V. F. Zolin, L. G. Koreneva, M. A. Samokhina, and V. F. Sodova, *Zh. Prikl. Spektrosk.* **20** (3), 516 (1974) [*Chem. Abstr.* **80**, 150879j].
26. B. L. Davydov, V. F. Zolin, L. G. Kureneva, and N. A. Lavrovskii, *Opt. Spectrosc. (Engl. Transl.)* **39** (4), 403 (1975).
27. V. D. Shigorin and G. P. Shipulo, *Sov. Phys.–Crystallogr. (Engl. Transl.)* **19** (5), 622 (1975).
28. J. R. Owen and E. A. D. White, *J. Mater. Sci.* **11**, 2165 (1976).
29. V. D. Shigorin, G. P. Shipulo, S. S. Grazhulene, L. A. Musikhin, and V. Sh. Shektman, *Sov. J. Quantum Electron. (Engl. Transl.)* **5** (11), 1393 (1976).
30. J. L. Oudar and R. Hierle, *J. Appl. Phys.* **48** (7), 2699 (1977).
31. A. Carenco, J. Jerphagnon, and A. Perigaud, *J. Chem. Phys.* **66** (8), 3806 (1977).
32. B. L. Davydov, S. G. Kotovshchikov, and V. A. Nefedov, *Sov. J. Quantum Electron (Engl. Transl.)* **7** (1), 129 (1977).
33. J. G. Bergman, J. Jerphagnon, and M. Perrin, *Chem. Phys. Lett.* **49** (2), 324 (1977).
34. O. S. Filipenko, V. D. Shigorin, V. I. Ponomarev, L. O. Atovmyan, Z. S. Safina, and B. L. Tarnopol'skii, *Sov. Phys.–Crystallogr. (Engl. Transl.)* **22** (3), 305 (1977).

35. D. Bäuerlé, K. Betzler, H. Hesse, S. Kapphan, and P. Loose, *Phys. Status Solidi A* **42,** K119 (1977).

36. J. G. Bergman and G. R. Crane, *J. Chem. Phys.* **66** (8), 3803 (1977).

37. E. M. Averyanov and V. F. Shabanov, *Opt. Spectrosc. (Engl. Transl.)* **44** (4), 410 (1978).

38. K. Betzler, H. Hesse, and P. Loose, *J. Mol. Struct.* **47,** 393 (1978).

39. B. F. Levine, C. G. Bethea, C. D. Thurmond, R. T. Lynch, and J. L. Bernstein, *J. Appl. Phys.* **50** (4), 2523 (1979).

40. C. Cassidy, J. M. Halbout, W. Donaldson, and C. L. Tang, *Opt. Commun.* **29** (2), 243 (1979).

41. J. M. Halbout, S. Blit, W. Donaldson, and C. L. Tang, *IEEE J. Quantum Electron.* **QE-15** (10), 1176 (1979).

42. M. Delfino, *Mol. Cryst. Liq. Cryst.* **52,** 271 (1979).

43. L. M. Belyaev, L. M. Dorozhkin, L. V. Soboleva, B. A. Chayanov, V. D. Shigorin, and G. P. Shipulo, *Sov. Phys.—Crystallogr. (Engl. Transl.)* **24** (4), 484 (1979).

44. J. M. Halbout, A. Sarhangi, and C. L. Tang, *Appl. Phys. Lett.* **37** (10), 864 (1980).

45. K. Kato, *IEEE J. Quantum Electron.* **QE-16** (8), 810 (1980).

46. O. S. Filipenko, V. I. Ponomarev, and L. O. Atovmyan, *Sov. Phys.—Crystallogr. (Engl. Transl.)* **25** (5), 549 (1980).

47. A. F. Garito, K. D. Singer, K. Hayes, G. F. Lipscomb, S. J. Lalama, and K. Desai, *J. Opt. Soc. Am.* **70** (11), 1399 (1980).

48. K. Kato, *IEEE J. Quantum Electron.* **QE-16** (12), 1288 (1980).

49. J. Zyss, D. S. Chemla, and J. F. Nicoud, *J. Chem. Phys.* **74** (9), 4800 (1981).

50. G. F. Lipscomb, A. F. Garito, and R. S. Narang, *Appl. Phys. Lett.* **38** (9), 663 (1981).

51. G. F. Lipscomb, A. F. Garito, and R. S. Narang, *J. Chem. Phys.* **75** (3), 1509 (1981).

52. J. M. Halbout, S. Blit, and C. L. Tang, *IEEE J. Quantum Electron.* **QE-17,** 513 (1981).

53. K. Jain, G. H. Hewig, Y. Y. Cheng, and J. I. Crowley, *IEEE J. Quantum Electron.* **QE-17** (9), 1593 (1981).

54. K. Jain, J. I. Crowley, G. H. Hewig, Y. Y. Cheng, and R. J. Twieg, *Opt. Laser Technol.* p. 297 (1981).

55. R. J. Twieg, K. Jain, Y. Y. Cheng, J. I. Crowley, and A. Azema, *Am. Chem. Soc., Div. Polym. Chem.* **23** (2), 147 (1982).

56. G. R. Meredith, *Polym. Prepr., Am. Chem. Soc., Div. Polym. Chem.* **23** (2), 158 (1982).

57. R. Twieg, A. Azema, K. Jain, and Y. Y. Cheng, *Chem. Phys. Lett.* **92** (2), 208 (1982).

58. J. Zyss, *J. Non-Cryst. Solids* **47** (2), 211 (1982).

59. J. M. Halbout and C. L. Tang, *IEEE J. Quantum Electron.* **QE-18,** 410 (1982).

60. A. F. Garito and K. D. Singer, *Laser Focus Fiberopt. Commun.* **18**(2), 59 (1982).

61. L. Schüler, K. Betzler, H. Hesse, and S. Kapphan, *Opt. Commun.* **43** (2), 157 (1982).

62. R. V. Vizgert, B. L. Davydov, S. G. Kotovshchikov, and M. P. Starodubsteva, *Sov. J. Quantum Electron. (Engl. Transl.)* **12** (2), 214 (1982).

63. M. Sigelle, J. Zyss, and R. Hierle, *J. Non-Cryst. Solids* **47,** 287 (1982).

64. D. Xu, M. Jiang, and Z. Tan, *Acta Chim. Sin.* **2,** 230 (1983).

65. A. E. Andreichuk, L. M. Durozhkin, Yu. I. Krasilov, I. A. Maslyanitsyn, S. M. Portnova, L. V. Soboleva, L. I. Khapaeva, B. A. Chayanov, V. D. Shigorin, and G. P. Shipulo, *Sov. Phys.—Crystallogr. (Engl. Transl.)* **28** (5), 547 (1983).

66. M. R. Gasparyan, A. V. Karmenyan, A. A. Martirosyan, A. M. Khachaturyan, and R. O. Sharkhatunyan, *Bull. Acad. Sci. USSR, Phys. Ser. (Engl. Transl.)* **47** (8), 130 (1983).

67. M. Perrin, A. Thozet, S. Lecocq, R. Perrin, and R. Lamartine, *Proc. SPIE Int. Soc. Opt. Eng.* **400,** 176 (1983).

68. M. J. Rosker and C. L. Tang, *IEEE J. Quantum Electron.* **QE-20** (4), 334 (1984).

69. A. F. Garito, C. C. Teng, K. Y. Wong, and O. Zammani'Khamiri, *Mol. Cryst. Liq. Cryst.* **106** (3, 4), 219 (1984).

70. N. G. Furmanova, Z. P. Razmanova, L. V. Soboleva, I. A. Maslyanitsyn, H. Siegert, V. D. Shigorin, and G. P. Shipulo, *Sov. Phys.—Crystallogr. (Engl. Transl.)* **29** (3), 285 (1984).

71. S. Tomaru, S. Zembutsu, M. Kawachi, and M. Kobayashi, *J. Chem. Soc., Chem. Commun.,* p 1207 (1984).

72. J. Zyss, J. F. Nicoud, and M. Coquillay, *J. Chem. Phys.* **81** (9), 4160 (1984).

73. G. R. Meredith, R. J. Weagley, D. J. Williams, and R. F. Ziolo, personal communication.

74. S. Basu, *Ind. Eng. Chem. Prod. Res. Dev.* **23** (2), 183 (1984).

75. D. J. Williams, *Angew. Chem., Int. Ed. Engl.* **23,** 690 (1984).

76. L. G. Koreneva, V. F. Zolin, and B. L. Davydov, "Molecular Crystals in Nonlinear Optics." Nauka, Moscow, 1975.

77. D. J. Williams, ed., "Nonlinear Optical Properties of Organic and Polymeric Materials," ACS Symp. Ser. No. 233. Am. Chem. Soc., Washington, D.C., 1983.

78. E. A. Tikhonov and M. T. Shpak, "Nonlinear Optical Phenomena in Organic Compounds." Naukova Dumka, Kiev, 1979 (not available for inclusion).

79. J. L. Oudar, Ph.D. Thesis, Université Paris VI, Paris (1977).

80. V. D. Shigorin, Ph.D. Dissertation, Phys. Inst. Akad. Nauk SSSR, Moscow (1976).

81. R. Bechmann and S. K. Kurtz, *in* "Landolt–Börnstein Tables," Vol. III, Part 2, p. 167. Springer–Verlag, Berlin and New York, 1969.

82. S. K. Kurtz, J. Jerphagnon, and M. M. Choy, *in* "Landolt–Börnstein Tables," Vol. III, Part 11, p. 671. Springer–Verlag, Berlin and New York 1979.

83. R. Twieg, unpublished results.

84. J. F. Nicoud, unpublished results.

85. R. Twieg and K. Jain, in Ref. 77, Ch. 3. *ACS Symp. Ser.* **233,** 57 (1983).

86. R. J. Twieg and N. E. Schlotter, *IBM Technical Disclosure Bulletin* **27(3),** 1538 (1984).

87. R. J. Twieg, D. Dobrowolski, and C. Dirk, unpublished results.

88. R. J. Twieg and C. Dirk, unpublished results.

89. R. J. Twieg and Y. Y. Cheng, unpublished results.

90. R. J. Twieg, E. Moret, and K. Jain, *IBM Technical Disclosure Bulletin* **26(1),** 422 (1983).

91. F. Pandarese, S. Panizza, and A. Telo, *Opt. Commun.* **12** (1), 14 (1974).

92. F. Bertinelli and C. Taliani, *Chem. Phys. Lett.* **28** (2), 231 (1974).

93. J. P. Dougherty and S. K. Kurtz, *J. Appl. Cryst.* **9,** 145 (1976).

94. A. Coda and F. Pandarese, *J. Appl. Cryst.* **9,** 193 (1976).

95. D. W. G. Ballentyne and S. M. Al-Shukri, *J. Cryst. Growth* **68,** 651 (1984).

96. R. S. Feigelson, R. K. Route, and T.-M. Kao, *J. Cryst. Growth* **72,** 585 (1985).

97. J. Jerphagnon, S. K. Kurtz, and J. L. Oudar, *in* "Landolt–Börnstein Tables," Vol. 18, Springer–Verlag. Berlin and New York, 1984.

Appendix **II**

Organic EFISH
Hyperpolarizability Data

J. F. NICOUD and R. J. TWIEG

This appendix is a compilation of hyperpolarizability data on π-bond-containing organic molecules obtained by the electric-field-induced second-harmonic (EFISH) technique (also known as dc SHG technique). The primary references (1–22) that contain the tabulated data are presented in chronological order (1974–1983). This appendix does not contain hyperpolarizability data that has been obtained in the gas phase (23–25) or in solution (2) for some relatively simple molecules. It is organized as a function of molecular functionality, substitution pattern, and extent of conjugation. Given the relatively small absolute number and the few structural types of molecules that have been measured by the EFISH technique, this organization, although somewhat arbitrary, is clear-cut.

Accurately determined solution EFISH data are extremely valuable to provide insight into the molecular structural and functional origins of hyperpolarizability. In solution the molecules are statistically isotropic but can be aligned by an external field. However, in a crystal there exist orientational influences involving the anisotropic dielectric media and optical phase-matching dependencies that significantly complicate the extraction of molecular properties out of the bulk hyperpolarizability properties. The role

255

of any solvent and dispersion effects on the EFISH measurements must be taken into account for any meaningful comparison among these materials. For this reason, the solvent and wavelength of the laser source utilized for the determination are provided whenever possible. In some of the references the issues of solvent effects (11, 14, 20) and dispersion (14, 18, 22) are specifically dealt with in some detail.

The experimental apparatus and methodology for the EFISH determination, such as cell design, are described in many of the primary references (2–4, 10, 12, 22) and in some secondary references (26, 27).

TABLE I

Mononuclear Aromatic Derivatives

A. Monosubstituted Benzene C_6H_5R

R	Name	$\|\beta\| \, (\times 10^{-40} \, m^4 V^{-1})^{a,b}$	Solvent	$\lambda \, (\mu m)$	Ref.	Miscellaneous $(\times 10^{-31} \, esu)$
NO_2	Nitrobenzene	8.4	benzene	1.06	1	20
		8.5	neat	1.06	2	−20.4
		9.6	neat	1.06	3	−23
		8.2	neat	1.318	4	19.7
		9.2	neat	1.06	8	22
		9.5	neat	1.06	9	22.7
CH_3	Toluene	0.8	neat	1.06	2	1.8
F	Fluorobenzene	1.7	neat	1.06	2	4.4
		2.2	neat	1.06	3	5.3
		2.9	neat	1.06	9	7.0
Cl	Chlorobenzene	1.2	neat	1.06	2	2.81
		0.9	neat	1.06	3	2.2
		1.4	neat	1.06	9	3.3
Br	Bromobenzene	0.2	neat	1.06	2	0.42
		0.1	neat	1.06	3	0.20
		0.8	neat	1.06	9	2.0
I	Iodobenzene	1.17	neat	1.06	2	−2.79
		2.93	neat	1.06	3	−7
		1.93	neat	1.06	9	4.6
NH_2	Aniline	3.72	neat	1.06	2	8.9
		5.15	neat	1.06	3	12.3
		3.31	neat	1.318	4	7.9
		4.61	neat	1.06	8	11.0
		6.20	neat	1.06	9	14.8
NMe_2	N,N-Dimethylaniline	7.33	neat	1.06	3	17.5
		6.74	neat	1.06	9	15.1
OH	Phenol	0.71	neat	1.06	3	−1.7
		1.61	neat	1.06	9	3.6

(continues)

TABLE I (*Continued*)

B. Orthodisubstituted Benzene 1-R-2-R′-C_6H_4

R	R′	Name	β ($\times 10^{-40}$ m^{4}V^{-1})	Solvent	λ (μm)	Ref.	Miscellaneous ($\times 10^{-31}$ esu)
F	NO_2	*o*-fluoronitrobenzene	7.33	neat	1.06	2	-17.5
NH_2	NO_2	*o*-nitroaniline	26.8	melt	1.318	4	64.0
			42.7	acetone	1.06	8	102.0

C. Metadisubstituted Benzene 1-R-3-R′-C_6H_4

R	R′	Name	β ($\times 10^{-40}$ m^{4}V^{-1})	Solvent	λ (μm)	Ref.	Miscellaneous ($\times 10^{-31}$ esu)
F	NO_2	*m*-Fluoronitrobenzene	6.8	neat	1.06	2	-16.4
NH_2	NO_2	*m*-Nitroaniline (mNA)	17.6	melt	1.318	4	42
			25.1	acetone	1.06	8	60
			25.5	acetone	1.06	9	61
NO_2	NO_2	*m*-Dinitrobenzene	5.0	benzene	1.06	9	12
NO_2	Cl	*m*-Chloronitrobenzene	6.7	benzene	1.06	9	16
NO_2	Br	*m*-Bromonitrobenzene	5.0	benzene	1.06	9	12
OH	OH	Resorcinol	0.8	ethanol	1.06	9	2
OH	NH_2	*m*-Aminophenol	5.0	acetone	1.06	9	12

D. Paradisubstituted Benzene 1-R-4-R'-C_6H_4

R	R'	Name	$\beta\,(\times 10^{-40}\,\mathrm{m^4 V^{-1}})$	Solvent	$\lambda\,(\mu m)$	Ref.	Miscellaneous
F	NO_2	*p*-Fluoronitrobenzene	8.96	neat	1.06	2	-21.4×10^{-31} esu
NH_2	NO_2	*p*-Nitroaniline (pNA)	151.0	methanol	1.06	3	360×10^{-31} esu
			88.3	melt	1.318	4	211×10^{-31} esu
			145.0	methanol	1.06	8	345×10^{-31} esu
			192–199	DMSO	1.89	12	28.6×10^{-47} $(\mu_0\beta)$ $\mu_0 = 6.2{-}6.0\,\mathrm{D}$
			20.1	stilbene	1.318	14	64×10^{-31} esu
			40.2	dioxane	1.907	22	96×10^{-31} esu
			49.4	dioxane	1.37	22	118×10^{-31} esu
			71.0	dioxane	1.06	22	16.9×10^{-31} esu
			105.0	dioxane	0.91	22	250×10^{-31} esu
			168.0	dioxane	0.83	22	400×10^{-31} esu
NMe_2	NO_2	*p*-*N*,*N*-Dimethylamino-nitrobenzene	215	DMSO	1.89	12	35.5×10^{-47} $(\mu_0\beta)$ $\mu_0 = 6.9\,\mathrm{D}$
NH_2	CN	*p*-Aminobenzonitrile	56	DMSO	1.89	12	7.9×10^{-47} $(\mu_0\beta)$ $\mu_0 = 5.92\,\mathrm{D}$
NMe_2	CN	*p*-*N*,*N*-Dimethylamino-benzonitrile	60	DMSO	1.89	12	9.4×10^{-47} $(\mu_0\beta)$ $\mu_0 = 6.6\,\mathrm{D}$
MeO	NO_2	*p*-Methoxynitrobenzene	59–73	DMSO	1.89	12	7.1×10^{-47} $(\mu_0\beta)$ $\mu_0 = 4.98{-}4.5\,\mathrm{D}$
Me	NO_2	*p*-Nitrotoluene	30.6–38.1	DMSO	1.89	12	3.3×10^{-47} $(\mu_0\beta)$ $\mu_0 = 4.5{-}3.62\,\mathrm{D}$
MeO	CN	*p*-Methoxybenzonitrile	20.1	DMSO	1.89	12	2.3×10^{-47} $(\mu_0\beta)$ $\mu_0 = 4.8\,\mathrm{D}$

(continues)

TABLE I (*Continued*)

R	R′	Name	β ($\times 10^{-40}$ m^{4}V^{-1})	Solvent	λ (μm)	Ref.	Miscellaneous
Me	CN	*p*-Tolunitrile	12	DMSO	1.89	12	1.3×10^{-47} ($\mu_0\beta$) $\mu_0 = 4.44$ D
NH$_2$	COCH$_3$	*p*-Aminoacetophenone	100	DMSO	1.89	12	10.9×10^{-47} ($\mu_0\beta$) $\mu_0 = 4.54$ D
OCH$_3$	CHO	*p*-Methoxybenzaldehyde	35[c]	DMSO	1.89	12	2.7×10^{-47} ($\mu_0\beta$) $\mu_0 = 3.26$ D
NMe$_2$	CHO	*p*-*N*,N-Dimethylamino-benzaldehyde	96[c]	DMSO	1.89	12	12.9×10^{-47} ($\mu_0\beta$) $\mu_0 = 5.6$ D

E. Trisubstituted Benzene

Name	β ($\times 10^{-40}$ m^{4}V^{-1})	Solvent	λ (μm)	Ref.	Miscellaneous ($\times 10^{-31}$ esu)
2,4-Dinitroaniline	88.0	acetone	1.06	8	210
Dinitrophenylmethylalaninate (DPMA, MAP)	92.0	acetone	1.06	8	220
2-Methyl-4-nitroaniline (MNA)	39.8	dioxane	1.907	22	95
	53.6	dioxane	1.37	22	128
	69.9	dioxane	1.06	22	167
	113.0	dioxane	0.91	22	270
	188.0	dioxane	0.83	22	450

TABLE II

R,R′-Substituted Aromatics with Extended Conjugation

R	R′	Name	$\beta\ (\times 10^{-40}\ \mathrm{m^4 V^{-1}})$	Solvent	$\lambda\ (\mu\mathrm{m})$	Ref.	Miscellaneous
		A. 4-R, β-R′ Styrene (K)					
Me$_2$N	NO$_2$	4-N,N-Dimethylamino-β-nitrostyrene (DMA-NS)	920 ± 160	CHCl$_3$	1.06	10	$2200 \pm 400 \times 10^{-31}$ esu
OCH$_3$	CHO	4-Methoxycinnamaldehyde	$52\text{–}62^c$	DMSO	1.89	17	$8 \times 10^{-47}\ (\mu_0\beta)$ $\mu_0 = 6.43\text{–}5.4$ D
Me$_2$N	CHO	4-N,N-Dimethylamino-cinnamaldehyde	—	DMSO	1.89	17	$37 \times 10^{-47}\ (\mu_0\beta)$
		B. α-(4-R-Phenyl)-δ-R′-butadiene					
Me$_2$N	NO$_2$	α-(4-N,N-Dimethylaminophenyl)-δ-nitrobutadiene (DMA-PNB)	2640 ± 670	acetone	1.06	10	$6300 \pm 1600 \times 10^{-31}$ esu
		C. α-(4-R-Phenyl)-δ-(4-R′-phenyl)butadiene (M)					
Me$_2$N	CN	α-(4-N,N-Dimethylaminophenyl)-δ-(4-cyanophenyl)butadiene	—	DMSO	1.89	17	$170 \times 10^{-47}\ (\mu_0\beta)$

(continues)

TABLE II (*Continued*)

R	R′	Name	$\beta\,(\times 10^{-40}\,\mathrm{m^4 V^{-1}})$	Solvent	$\lambda\,(\mu m)$	Ref.	Miscellaneous
		D. 4-R-4′-R′-Stilbene (L)					
NMe_2	NO_2	4-N,N-Dimethylamino-4′-nitrostilbene (DANS)	1900	acetone	1.06	3	4500×10^{-31} esu
			1900 ± 380	acetone	1.06	10	$4500 \pm 90 \times 10^{-31}$ esu
			1633 ± 245	stilbene	1.318	14	3900×10^{-31} esu
			$2478 \pm 20\%^{c}$	DMSO	1.89	17	$420 \times 10^{-47}\,(\mu_0\beta)$ $\mu_0 = 7.1$ D
NO_2	H	4-Nitrostilbene	121 ± 49	benzene	1.06	10	$290 \pm 100 \times 10^{-31}$ esu
NH_2	H	4-Aminostilbene	50 ± 29	benzene	1.06	10	$120 \pm 70 \times 10^{-31}$ esu
NMe_2	H	4-N,N-Dimethylaminostilbene	121 ± 33	benzene	1.06	10	$290 \pm 80 \times 10^{-31}$ esu
Cl	H	4-Chlorostilbene	-151 ± 84	benzene	1.06	10	$-36 \pm 20 \times 10^{-31}$ esu
Cl	NO_2	4-Chloro-4′-nitrostilbene	163 ± 41	$CHCl_3$	1.06	10	$390 \pm 100 \times 10^{-31}$ esu
NMe_2	Cl	4-N,N-Dimethylamino-4′-chlorostilbene	176 ± 41	$CHCl_3$	1.06	10	$420 \pm 100 \times 10^{-31}$ esu
NH_2	NO_2	4′-Amino-4′-nitrostilbene	1100 ± 150	acetone	1.06	10	$2600 \pm 350 \times 10^{-31}$ esu
NMe_2	CN	4-N,N-Dimethylamino-4′-cyanostilbene	$483\,(\pm 20\%)$	DMSO	1.89	17	$82 \times 10^{-47}\,(\mu_0\beta)$ $\mu_0 = 7.1$ D
OCH_3	CN	4-Methoxy-4′-cyanostilbene	—	DMSO	1.89	17	$9.8 \times 10^{-47}\,(\mu_0\beta)$
CH_3	NO_2	4-Methyl-4′-nitrostilbene	—	DMSO	1.89	17	$20 \times 10^{-47}\,(\mu_0\beta)$

TABLE III

Cyanine and Merocyanine Dyes

Name	$\beta\,(\times 10^{-40}\,\mathrm{m^4V^{-1}})$	Solvent	$\lambda\,(\mu m)$	Ref.	Miscellaneous
N-Octadecylmerocyanine	$-418\,(-214)^d$	MeOH	1.318	11	$-1000\,(-510)^d \times 10^{-31}$ esu
	$-543\,(-155)$	1% 1 MeOH/pyridine	1.318	11	$-1300\,(-370) \times 10^{-31}$ esu
	$-880\,(-142)$	pyridine	1.318	11	$-2100\,(-340) \times 10^{-31}$ esu
N-Methylmerocyanine	4000	DMSO	1.89	13	$-7.6 \times 10^{-45}\,(\mu_0\beta)$ and $\mu \approx 8 \times 10^{-18}$ esu
	$(1200)^e$	DMSO	1.89		using $\mu_0 = 26\,\mathrm{D}^e$
Miscellaneous cyanines	—	DMSO	1.89	15	$\mu_0\beta$ values only

TABLE IV

Pyridine *N*-Oxide Derivatives

Name	$\beta\,(\times 10^{-40}\ \mathrm{m^4 V^{-1}})$	Solvent	$\lambda\,(\mu\mathrm{m})$	Ref.	Miscellaneous
3-Methyl-4-nitropyridine *N*-oxide (POM)	35.6	Crystal data	1.06	21	$85 \pm 20 \times 10^{-31}$ esu
4-*N*,*N*-Dimethylaminopyridine *N*-oxide (DMAPO)	21	$CHCl_3$	1.06	19	50×10^{-31} esu
4'-*N*,*N*-Dimethylamino-4'-azastilbene *N*-oxide	2090	$CHCl_3$	1.06	19	5000×10^{-31} esu

TABLE V

Urea and Derivatives

Name	β ($\times 10^{-40}$ m^{4}V^{-1})	Solvent	λ (μm)	Ref.	Miscellaneous ($\times 10^{-31}$ esu)
Urea	9.6	H_2O	1.06	16	23
	1.9	H_2O	1.06	20	4.5 ± 1.2
	1.3	DMF	1.06	20	3 ± 0.9
	1.2	DMSO	1.06	20	2.9 ± 0.3
Monomethylurea	1.9	H_2O	1.06	18	4.5 ± 1.6
	2.7	DMF	1.06	18	6.5 ± 1.2
		DMSO	1.06	18	4.3 ± 0.4
sym-Dimethylurea	2.5	H_2O	1.06	18	6 ± 3
	1.2	DMF	1.06	18	2.8 ± 2.0
	0.8	DMSO	1.06	18	1.8 ± 1.3
Tetramethylurea	1.7	Neat	1.06	18	4.1 ± 0.2

TABLE VI

Miscellaneous Compounds

Name	$\beta\,(\times 10^{-40}\ \mathrm{m^4V^{-1}})$	Solvent	$\lambda\,(\mu\mathrm{m})$	Ref.	Miscellaneous
Pyridine-I_2 CT	$40 \pm 15\%$	benzene	1.318	6	96×10^{-31} esu $\pm 15\%$
Pyridine-ICl CT	$58 \pm 15\%$	9:1 benzene/pyridine	1.318	7	45×10^{-31} esu
4-Aminopyridine-I_2 CT	$18.8 \pm 15\%$	dioxane	1.318	7	45×10^{-31} esu
2-(4-Dicyanomethylene cyclohexa-2,5-dienylidene) imidazolidine (DCNQI)	-1000 ± 250	DMSO	1.06	18	-2400 ± 600 $\times 10^{-31}$ esu
Poly-γ-benzyl-L-glutamate (PBLG), MW 550K	$2090 \pm 50\%$	dichloroethylene	1.06	5	5000×10^{-31} esu $\mu = 8000$ D

[a] For monosubstituted benzene derivatives, only $|\beta|$ is provided. See original references for sign conventions.

[b] β (MKS in $10^{-40}\ \mathrm{m^4V^{-1}}$) $= 4.189 \times 10^{-10}\ \beta$ (CGS in esu).

[c] In a few cases, additional μ_0 data was located (A. L. McClellan, "Tables of Experimental Dipole Moments." Freeman, San Francisco, California, 1963) to allow calculation of β from Dulcic's $\mu_0\beta$ data (refs. 12, 13, 15, 17). In cases where the range of μ_0 is large, a range of β is provided otherwise an average value is provided. The conversion is β (MKS in $10^{-40}\ \mathrm{m^4V^{-1}}$) $= 4.189 \times 10^8 \times \mu_0\beta$ (Dulcic)$/\mu_0$ where μ_0 is in Debye units.

[d] Values in parentheses are corrected for dispersion.

[e] The original reference 13 underestimates the μ_0 of merocyanine.

REFERENCES

1. B. F. Levine and C. G. Bethea, *Appl. Phys. Lett.* **24**(9), 445 (1974).
2. B. F. Levine and C. G. Bethea, *J. Chem. Phys.* **63**(6), 2666 (1975).
3. J. L. Oudar and H. Le Person, *Opt. Commun.* **15**(2), 258 (1975), errata: *ibid.* **18**(3), 410 (1976).
4. B. F. Levine, *Chem. Phys. Lett.* **37**(3), 516 (1976).
5. B. F. Levine and C. G. Bethea, *J. Chem. Phys.* **65**(5), 1989 (1976).
6. B. F. Levine and C. G. Bethea, *J. Chem. Phys.* **65**(6), 2439 (1976).
7. B. F. Levine and C. G. Bethea, *J. Chem. Phys.* **66**(3), 1070 (1977).
8. J. L. Oudar and D. S. Chemla, *J. Chem. Phys.* **66**(6), 2664 (1977).
9. J. L. Oudar, D. S. Chemla, and E. Batifol, *J. Chem. Phys.* **67**(4), 1626 (1977).
10. J. L. Oudar, *J. Chem. Phys.* **67**(2), 446 (1977).
11. B. F. Levine, C. G. Bethea, E. Wasserman, and L. Leenders, *J. Chem. Phys.* **68**(11), 5042 (1978).
12. A. Dulcic and C. Sauteret, *J. Chem. Phys.* **69**(8), 3453 (1978).
13. A. Dulcic and C. Flytzanis, *Opt. Commun.* **25**(3), 402 (1978).
14. B. F. Levine and C. G. Bethea, *J. Chem. Phys.* **69**(12), 5240 (1978).
15. A. Dulcic, *Chem. Phys.* **37**, 57 (1979).
16. C. Cassidy, J. M. Halbout, W. Donaldson, and C. L. Tang, *Opt. Commun.* **29**(2), 243 (1977).
17. A. Dulcic, C. Flytzanis, C. L. Tang, D. Pépin, M. Fétizon, and Y. Hoppilliard, *J. Chem. Phys.* **74**(3), 1559 (1981).
18. S. J. Lalama, K. D. Singer, A. F. Garito, and K. N. Desai, *Appl. Phys. Lett.* **39**(12), 940 (1981).
19. J. Zyss, D. S. Chemla, and J. F. Nicoud, *J. Chem. Phys.* **74**(9), 4800 (1981).
20. I. Ledoux and J. Zyss, *Chem. Phys.* **73**, 203 (1982).
21. J. Zyss and J. L. Oudar, *Phys. Rev. A* **26**(4), 2028 (1982).
22. C. C. Teng and A. F. Garito, *Phys. Rev. B* **28**(12), 6766 (1983).
23. G. Mayer, *C. R. Hebd. Seances Acad. Sci., Ser. B* **267**, 54 (1968).
24. G. Hauchecorne, F. Kerhervé, and G. Mayer, *J. Phys. Orsay, Fr.* **32**, 47 (1971).
25. J. F. Ward and C. K. Miller, *Phys, Rev. A* **19**, 826 (1979).
26. G. R. Meredith, *Rev. Sci. Instrum.* **53**, 48 (1982).
27. D. J. Williams, *Agnew. Chem., Int. Ed. Engl.* **23**, 690 (1984).

Index

T